Modeling Operations Research and Business Analytics

This book provides sample exercises, techniques, and solutions to employ mathematical modeling to solve problems in Operations Research and Business Analytics. Each chapter begins with a scenario and includes exercises built on realistic problems faced by managers and others working in operations research, business analytics, and other fields employing applied mathematics. A set of assumptions is presented, and then a model is formulated. A solution is offered, followed by examples of how that model can be used to address related issues.

Key elements of this book include the most common problems the authors have encountered over research and while consulting the fields including inventory theory, facilities' location, linear and integer programming, assignment, transportation and shipping, critical path, dynamic programming, queuing models, simulation models, reliability of system, multi-attribute decision-making, and game theory.

In the hands of an experienced professional, mathematical modeling can be a powerful tool. This book presents situations and models to help both professionals and students learn to employ these techniques to improve outcomes and to make addressing real business problems easier. The book is essential for all managers and others who would use mathematics to improve their problem-solving techniques.

No previous exposure to mathematical modeling is required. The book can then be used for a first course on modeling, or by those with more experience who want to refresh their memories when they find themselves facing real-world problems. The problems chosen are presented to represent those faced by practitioners.

The authors have been teaching mathematical modeling to students and professionals for nearly 40 years. This book is presented to offer their experience and techniques to instructors, students, and professionals.

Advances in Applied Mathematics

Series Editor: Daniel Zwillinger

Advanced Engineering Mathematics
A Second Course with MATLAB®
Dean G. Duffy

Quantum Computation
Helmut Bez and Tony Croft

Computational Mathematics
An Introduction to Numerical Analysis and Scientific Computing with Python
Dimitrios Mitsotakis

Delay Ordinary and Partial Differential Equations
Andrei D. Polyanin, Vsevolod G. Sorkin and Alexi I. Zhurov

Clean Numerical Simulation
Shijun Liao

Multiplicative Partial Differential Equations
Svetlin Georgiev and Khaled Zennir

Engineering Statistics
A Matrix-Vector Approach with MATLAB®
Lester W. Schmerr Jr.

General Quantum Numerical Analysis
Svetlin Georgiev and Khaled Zennir

An Introduction to Partial Differential Equations with MATLAB®
Matthew P. Coleman and Vladislav Bukshtynov

Handbook of Exact Solutions to Mathematical Equations
Andrei D. Polyanin

Introducing Game Theory and its Applications, Second Edition
Elliott Mendleson and Dan Zwillinger

Modeling Operations Research and Business Analytics
William P. Fox and Robert E. Burks

*www.routledge.com/Advances-in-Applied-Mathematics/book-series/CRCA
DVAPPMTH?pd=published,forthcoming&pg=1&pp=12&so=pub&view=list*

Modeling Operations Research and Business Analytics

William P. Fox and Robert E. Burks

CRC Press
Taylor & Francis Group
Boca Raton London New York

CRC Press is an imprint of the
Taylor & Francis Group, an **informa** business

A CHAPMAN & HALL BOOK

First edition published 2025
by CRC Press
2385 Executive Center Drive, Suite 320, Boca Raton, FL 33431

and by CRC Press
4 Park Square, Milton Park, Abingdon, Oxon, OX14 4RN

CRC Press is an imprint of Taylor & Francis Group, LLC

© 2025 William P. Fox and Robert E. Burks

ISBN: 978-1-032-71755-5 (hbk)
ISBN: 978-1-032-73592-4 (pbk)
ISBN: 978-1-003-46496-9 (ebk)

DOI: 10.1201/9781003464969

Typeset in Minion
by Apex CoVantage, LLC

Contents

About the Authors

William P. Fox (Ph.D.) is currently a Visiting Professor of Computational Operations Research at the College of William and Mary, Williamsburg, Virginia, United States. He is an Emeritus Professor in the Department of Defense Analysis at the Naval Postgraduate School and teaches a three-course sequence in mathematical modeling for decision-making. He received his Ph.D. in Industrial Engineering from Clemson University. He taught at the United States Military Academy for 12 years until retiring and at Francis Marion University where he was the chair of mathematics for eight years. He has many publications and scholarly activities including 20-plus books and 150 journal articles to his name.

Colonel (R) Robert E. Burks, Jr. (Ph.D.) is an Associate Professor in the Defense Analysis Department of the Naval Postgraduate School (NPS) and the Director of the NPS' Wargaming Center. He holds a Ph.D. in Operations Research from the Air Force Institute of Technology. He is a retired logistics Army Colonel with more than 30 years of military experience in leadership, advanced analytics, decision modeling, and logistics operations who served as an Army Operations Research Analyst at the Naval Postgraduate School, TRADOC Analysis Center, United States Military Academy, and the United States Army Recruiting Command.

Another book by William P. Fox and Robert E. Burks is *Advanced Mathematical Modeling with Technology*, 2021, CRC Press.

OTHER BOOKS BY WILLIAM P. FOX FROM CRC PRESS

Probability and Statistics for Engineering and the Sciences with Modeling Using R (w/Rodney X. Sturdivant, 2023), CRC Press.

Mathematical Modeling in the Age of the Pandemic, 2021, CRC Press.

Advanced Problem Solving Using Maple: Applied Mathematics, Operations Research, Business Analytics, and Decision Analysis (w/William Bauldry), 2020, CRC Press.

Mathematical Modeling with Excel (w/Brian Albright), 2020, CRC Press.

Nonlinear Optimization: Models and Applications, 2020, CRC Press.

Advanced Problem Solving with Maple: A First Course (w/William Bauldry), 2019, CRC Press.

Mathematical Modeling for Business Analytics, 2018, CRC Press.

Preface

AUDIENCE

Operations and Business Analytics is a craft that requires constant learning and practice for the practitioner to maintain and enhance their ability to use those skills and to continue to develop within the craft of research and analysis. Frankly, the more practice you get, the better you will become in the craft of operations research (OR) analysis. This book covers the topics of OR and business analysis that we teach or have taught over the past 30 years. Each chapter will begin with a problem that needs to be solved to help motivate the major focus of the chapter. The key focus is process and techniques to solve the model, and, in many cases, we will present the technology as a method to solve that model or perform more in-depth analysis. Over the past 15 years, the authors have taught multiple mathematical modeling courses to students where Excel was chosen as the software of choice because of its availability, ease of use, familiarity with students, and access in their future jobs and careers.

OBJECTIVES

The objective of *Modeling Operations Research and Business Analytics* is to illustrate advanced applied mathematical modeling techniques that are accessible to the practitioner from many disciplines. The goal is that this book is used as part of the craft of developing operations research and business analytics and support fostering a desire for lifelong mathematical learning, habits of mind, and preparing competent and confident problem solvers for the twenty-first century. We caution that this book is not intended to serve as a textbook for an introduction to operations research or mathematical modeling course. It is designed to be used for individuals who already have a foundational or introductory level of basic mathematical modeling knowledge and have been introduced to technology at some level. *Modeling Operations Research and Business Analytics* would be of interest to mathematics departments that offer courses focused on decision-making or discrete mathematical modeling, and to an individual looking for an opportunity to practice the craft of operations research.

TECHNOLOGY

Technology is just a means to an end. It is not and should not be the focus of any individual until they have a strong OR foundation. We have selected Excel and R to support, where applicable to the content, the mathematical modeling process because of their wide accessibility to most individuals. We believe it is fundamental that individuals have a basic foundational understanding of at least one of these technologies to get the most out of this book. We provide the appropriate R code for the relevant technologies that we think are applicable to solve a problem in each chapter. In this way, the chapter can be covered in general discussion and the technology chosen from those provided to illustrate the models. However, we do not provide a discussion covering the basic setup of the technology. Many of the figures in this book are generated with one of the two selected technologies. It is important to note that one limitation of Excel is its limited graphics capabilities, especially in three dimensions. Although we attempt to use all features available in Excel, occasionally, we felt the need to create Excel templates, macros, and programs to solve the problem. This material is available upon request.

ORGANIZATION

This book contains information that could easily be covered in an advanced semester course focused on operations research and mathematical modeling or a semester-long survey course of the various topics in the book. The book is designed to provide instructors the flexibility to pick and select material to support their course.

- Chapters 1 and 2 provide a good foundation for understanding the common business-related problems of inventory, transportation, facility locations, and product mix. The two chapters cover important concepts from economic order quantity and the related time variant to product mix and shipping. Chapter 2 will cover the use of linear programming to solve common economy-related problems that frequently occur in modern industry. Areas using linear programming are as diverse as defense, health, transportation, manufacturing, advertising, and telecommunications. The reason linear programming is so popular is the classic economic problem of wanting to maximize output with limited resources. The chapter provides multiple techniques to handle these problems and the importance of conducting sensitivity analysis to better understand the impact of the developed solutions. In addition, each chapter provides technology examples to determine and understand the behavior of these systems.

- Chapter 3 builds upon the concepts and knowledge gained or refreshed in the previous chapters. It will expand upon the transportation and shipping examples presented in Chapter 2 and is designed to introduce the reader to additional linear programming problems.

- Chapter 4 continues the linear programming theme and introduces the topic of assignment models. Assignment models are a special application of the linear programming theme, where the objective is to assign the work/task to a group or individuals. This is typically done to minimize or maximize some parameter of interest such as cost or profits. The reader will gain a better appreciation of the power of linear programming in addressing these types of problems after working through the problem examples in this chapter.

- Chapter 5 discusses several mathematical programming methods that include Data Envelopment Analysis (DEA), flow problems, critical path, and both general and mixed integer programming problems. The topics covered in this chapter get to one of the critical foundation skills of all operations research analysts – integer programming problems. Each of the methods presented in this chapter will support the OR practitioner in addressing a range of business-related problems.

- Chapter 6 covers the critical application of dynamic programming to solve both discrete and continuous nonlinear, multistage problems. This chapter will help the practitioner to understand the dynamic programming process of simplifying a decision by breaking it into a sequence of smaller decision steps and then using an optimization process to recursively solve each step. The chapter will take the practitioner through the basic theory of dynamic programming to modeling application of discrete dynamic programming.

- Chapter 7 covers queuing models and helps the practitioner to develop an understanding of queuing theory and its application in any organization that is providing a service. The single and multi-server applications will set the foundation for the reader to build upon in future chapters.

- Chapter 8 discusses the power and limitations of simulations in developing solutions to problems. The chapter builds upon the queuing theory covered in Chapter 7 and extends it to develop an understanding of the concept of algorithms while building both deterministic and stochastic simulations to solve problems.

- Chapter 9 addresses how to determine the reliability or expected failure time of common systems. The chapter demonstrates how we can use basic probability concepts to determine the reliability of equipment. The practitioner will develop models for system components that are in series or parallel to determine the reliability of the system. The chapter will also review how to determine the reliability of active and redundant systems.

- Chapter 10 discusses the use of multi-attribute decision-making (MADM) in solving problems. The real world is filled with problems that have multiple criteria to consider in weighing alternatives and courses of action. This chapter covers several weighting schemes to include entropy and pairwise comparison. The reader will also gain good understanding of data envelopment analysis, order preference by similarity, and sum of additive weights in the chapter.

- Chapter 11 covers the use of regression analysis tools such as linear regression and polynomial regression. The chapter provides an understanding of the concepts of correlation, covariance, and the diagnostics of regression analysis. It will use several examples to build and interpret both linear and nonlinear regression models.

- Chapter 12 covers the use of game theory in solving problems. The chapter covers both total and partial conflict games and the methods to obtain equalizing strategies. The chapter uses multiple case studies to demonstrate the use of game theory to show the type of real decision problems and analysis.

Unfortunately, the length of this book prevents us from addressing all the potential nuances involved in modeling real-world problems. We have attempted to provide a set of models and solution techniques to obtain useful results for many real-world common problems.

We thank all the mathematical modeling students that we have had over the last 30 years as well as all the colleagues who have taught mathematical modeling with us during this adventure. We particularly single out the following who helped in our three-course mathematical modeling sequence at the Naval Postgraduate School over the years: Bard Mansger, Mike Jaye, Steve Horton, Patrick Driscoll, and Greg Mislick. We are especially appreciative of the mentorship of Frank R. Giordano over the past 30-plus years.

William P. Fox
College of William and Mary

Robert E. Burks
Naval Postgraduate School

CHAPTER 1

Inventory Problem

1.1 INTRODUCTION

The theory of inventory and production management is a subspecialty within operations research and operations management. Typically, it is focused on the design of inventory systems to minimize costs or maximize profit. It is the study of decisions faced by businesses and the military in all aspects of operations to include manufacturing, warehousing, supply chains, and resource allocation and provides the mathematical foundation for logistics. Every organization faces some version of the inventory problem where decision-makers must determine how much to order and in what time period to meet the demands and requirements. The problem has a robust modeling foundation using mathematical techniques of optimal control, dynamic programming, and network optimization.

1.2 INVENTORY PROBLEMS

A starting example is given here (adapted from Giordano et al., 2014).

Consider that you are the manager of a chain of BP gas stations in Virginia, and as the manager you need to determine how often and how much gasoline your suppliers should deliver to your various stations. The gasoline stations are located near interstate highways, where demand is seen to be fairly constant throughout the week. Records indicating the number of gallons sold daily are available from each of your gas stations.

After some questioning and analyzing past deliveries, you determine that each time gasoline is delivered, stations incur a constant charge of d dollars. This cost is in addition to the cost of the gasoline and is independent of the amount of gasoline they actually received during the delivery.

DOI: 10.1201/9781003464969-1

In addition, each station has underground storage tanks. Costs are also incurred when the gasoline is stored in these tanks. There are two types of costs that you should consider here: Cost of storage (implying nonuse) and inventory and the amortization of the tanks, equipment, insurances, taxes, security, etc.

1.2.1 Problem of Interest

Assume as the manager that you desire to maximize profits and that demand and price are constant in the short run (over the next six months). Thus, since total revenue is constant, total profit can be maximized by minimizing the total costs. There are many other components of total costs such as overhead, employee costs, and the like, but we will initially ignore those in our model. We will assume for the moment that the costs are not so affected and focus our attention on the following problem statement: *Minimize the average daily cost of delivery and storing sufficient gasoline as the best inventory policy at each station to meet consumer demand.* Mathematically, we expect such a minimum to exist as we will later show.

1.2.2 Assumptions

We consider some important factors to decide how large an inventory to maintain. Inventory-carrying cost might include:

(1) Cost of money tied up in inventory.

(2) Cost of goods which might deteriorate over time while being stored in inventory.

(3) Cost of the storage space.

(4) Cost of obsolescence or failure to get each type of gasoline needed.

Ordering costs must might include:

(1) Cost of placing an order (if not a standing order).

(2) Cost of receiving and unloading an order.

(3) Cost of processing the order and payment.

Now consider for a moment the volatility of gasoline market prices in 2022–2023. The supplier might be reluctant to store large quantities of the product. Location and sale of electric cars in the region might also have an impact on the amount of gasoline required and demanded. Some owners would opt for a more costly inventory strategy to assure that they never run out of gas. From this short discussion, we can see that any inventory decision is not an easy one. To get started, we restrict our initial model here to the following variables:

Average daily cost = f(storage costs, delivery costs, demand rate)

1.2.3 Storage Costs

We need to consider how the storage cost per unit varies with the number of units being stored. Are we renting space and receiving a discount when storage exceeds certain levels, or do we rent the cheapest storage first (adding more space as needed)? Do we need to rent an entire warehouse or floor? If so, then the per unit price is likely to decrease as the quantity stored increases until another warehouse or floor needs to be rented. Does the company own its own storage facilities? If so, what alternative use can be made of them? These are all valid options. For simplicity, we will assume that per unit storage is a constant in our model.

1.2.4 Delivery Costs

In many cases, the delivery charge depends on the amount delivered. For example, if a larger truck or an extra flatcar is needed, an additional charge is made. In our model, we consider a constant delivery charge independent of the amount delivered. We assume the tankers bring the gasoline to the station and fill the storage tanks with the amount of gasoline ordered.

1.2.5 Demand

Even though we realize that the demands occur in discrete time periods, for purposes of simplification, we take a continuous model for demand. Our continuous model is a line through the origin with a positive slope, where the slope of the line represents the constant demand rate. Notice the importance of our assumptions in producing the linear model.

1.2.6 Model Formulation

We use the following notation for constructing our model:

s = Storage costs, per gallon, per day

d = Delivery cost, in dollars, per delivery

r = Demand rate, in gallons, per day

Q = Quantity of gasoline, in gallons

T = Time in days

Now suppose an amount of gasoline, say $Q = q$, is delivered at time $T = t$ and that the gasoline is used up after $T = t$ days. The same cycle is then repeated, as illustrated in the saw tooth graph. The slope of each line segment is $-r$ (the negative of the demand rate). The problem is to determine an order quantity Q^* and a time between order T^* that minimizes the delivery and storage costs. See Figure 1.1 for an illustration of this model.

We seek an expression for the average daily cost, so consider the delivery and storage costs for a cycle of length t days. The delivery costs are the constant amount, d, since only one delivery is made over the single time period. To compute the storage cost, we take the average daily inventory, $q/2$, multiply by the number of days in storage, t, and multiply that by the storage cost per item per day, s. In our notation, this gives:

Cost per Cycle $= d + s\,\dfrac{q}{2}t$, which upon division by t yields the average daily Cost:

$$c = \frac{d}{t} + \frac{sq}{2}$$

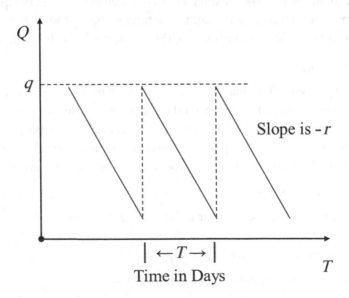

FIGURE 1.1   A Typical Inventory Cycle.

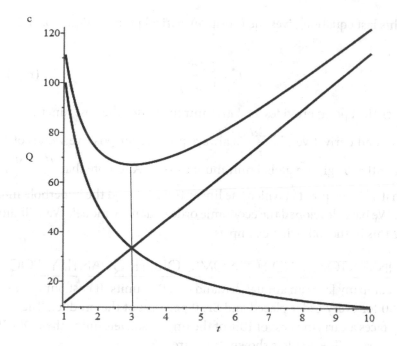

FIGURE 1.2 Typical Average Daily Cost Curve as a Sum of a Hyperbola and a Line.

1.2.7 Model Solution

Apparently, the cost function to be minimized has two independent variables – q and t.

For our single cyclic period, the amount delivered equals the amount demanded. This translates to $q = rt$. Substitution into the average daily cost equation yields:

$$c = \frac{d}{t} + \frac{srt}{2}$$ (Eq. 1.1)

Equation 1.1 is the sum of a hyperbola and a linear function. The situation is depicted in Figure 1.2.

As we see in Figure 1.2, we are guaranteed a local minimum that occurs at the intersection of the line and the hyperbola.

Let's find the time T^* between orders, which minimizes the average daily cost. Differentiating c with respect to t and setting $c' = 0$ yields

$$c' = -\frac{d}{t^2} + \frac{sr}{2} = 0.$$

This last equation gives the (positive) critical point as Equation 1.2:

$$T^* = \left(\frac{2d}{sr}\right)^{\!1/2}$$

(Eq. 1.2)

This critical point provides a relative minimum for the cost function since the second derivative $c'' = \dfrac{2d}{t^3}$ is always positive for positive values of t. It is clear that T^* gives a global minimum as well. And note that $\dfrac{d}{T^*} = \dfrac{sr}{2}T^*$, so that T^* is the point at which the linear function and the hyperbola intersect. We have developed the economic order quantity model. We will illustrate this in the following examples.

1.3 INVENTORY AND ECONOMIC ORDER QUANTITY (EOQ)

Our company faces an annual demand of 2,000 units. It cost our company $1,000 for every order placed and $250 per unit of the product. The company faces a carrying cost of 10% of the unit cost. Determine the EOQ. We know $q = r\,t$. The result is shown in Figure 1.3.

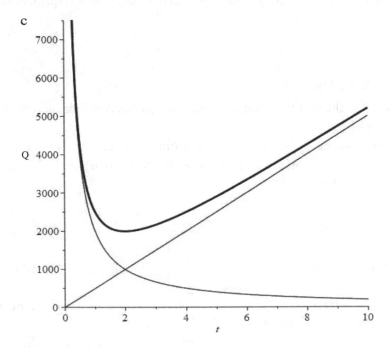

FIGURE 1.3 Graphs of Models of EOQ Model as the Sum of a Linear Function and a Hyperbola.

$d = 2,000$ units

$Q = q$

$S = 1,000$

$C = 250$

$H = \$25$

$I = 10\%$

$TC = 2000/t + 1000\ t\ (1)/2$

We might develop a spreadsheet to do this using our formulas. Table 1.1 provides the calculations from the formulas in Excel, and Figure 1.4 provides the graphical results.

1.3.1 Inventory Analysis with EOQ Formula-Driven Approach

From our previous analytical solutions in Section 1.2.7:

Let R be the uniform demand (units) per time period.

Let C_o be the initial setup cost in dollars.

Let C_i be the cost of carrying one unit of inventory for one time period.

Let q be the quantity (size) of manufactured lot.

The problem is to find the optimal value of q such that the overall cost is minimized.

Our costs are the sum of setup and inventory-carrying costs:

$$C_t = (R/q)\ C_o + (1/2)\ q\ C_i \qquad \text{(Eq. 1.3)}$$

As previously shown, calculus can be used to obtain the quantity to yield the minimum cost.

$$q^* = \sqrt{\frac{2RC_0}{C_i}} \qquad \text{(Eq.1.4)}$$

The corresponding number of setups is

$$S_o = R/q^* \qquad \text{(Eq. 1.5)}$$

TABLE 1.1   Excel Output of Model

		EOQ Excel		
	D		2000	
	S		1000	
	H		25	
Q	D/Q	D/Q*S	(Q/2)*H	TC
50	40,00	40,000,00	625.00	40,625.00
100	20.00	20,000.00	1,250.00	21,250.00
150	13.33	13,333.33	1.875.00	15,208.30
200	10.00	10,000.00	2.500.00	12,500.00
250	8.00	8,000.00	3,125.00	11,125.00
300	6.67	6,666.67	3,750.00	10,416.70
350	5.71	5,714,29	4,375.00	10,089.30
400	5.00	5,000.00	5.000.00	10,000.00
450	4,44	4,444.44	5,625.00	10,069.40
500	4,00	4,000.00	6,250.00	10,250.00
550	3.64	3,636.36	6.875.00	10,511.40
600	3.33	3,333.33	7,500.00	10,833.30
650	3.08	3,076.92	8,125.00	11,201.90
700	2.86	2,857.14	8,750.00	11,607.10
750	2.67	2,666.67	9.375.00	12,041,70
800	2.50	2,500.00	10.000.00	12,500.00
850	2.35	2,352.94	10,625.00	12,977.90
900	2.22	2,222.22	11,250.00	13,472.20
950	2.11	2,105.26	11.875.00	13,980.30
1000	2.00	2.000.00	12.500.00	14,500.00
1050	1.90	1,904.76	13,125.00	15,029.80
1100	1.82	1,818.18	13,750.00	15,568.20
1150	1.74	1,739.13	14,375.00	16,114.10

The corresponding overall minimum cost is

$$C_t^* = \sqrt{2RC_oC_i} \qquad \text{(Eq. 1.6)}$$

The number of setups will most likely not be an integer. In this case, the integer values are on each side of S_o. Then use the equation, $S_o = R/q^*$ to obtain the corresponding q^*. Then substitute into the formula to see which produces the lower total cost.

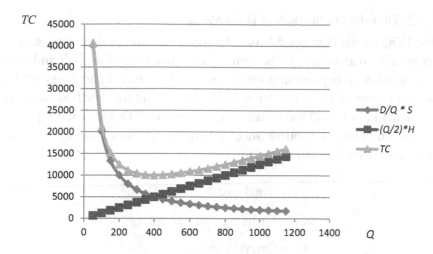

FIGURE 1.4 The Excel Plot of the Linear and Hyperbolic Functions as well as Total Cost Function from EOQ Model.

Example numbers for this method.

R = 50,000 units per year
C_o = $1,000 setup cost
C_i = $1 for carrying one unit of inventory for one year.

$$q^* = \sqrt{\frac{2RC_0}{C_i}}$$

and by substitution q^* = 10,000 units

$$S_o = 50{,}000/10{,}000 = 5 \text{ setups}$$

$C_t^* = \sqrt{2RC_oC_i}$ and substituting we find our minimized total cost is $10,000 or substituting into our original cost equation,

$$C_t = (R/q)\ C_o + (1/2)\ q\ C_i = 5 \times 1000$$
$$+ \tfrac{1}{2}\ (10{,}000)\ 1 = \$10{,}000 \tag{Eq. 1.7}$$

1.3.2 Time-Invariant Asphalt EOQ Model

The EOQ model developed is based upon an annual period of time. It is reasonable to assume that the demand of items, such as asphalt and perhaps gasoline, is dependent upon the season the year. Let's assume that the demand from October to March is 9,000 tons and for April–September, the demand is 15,000 tons. Can we calculate our EOQ for each separate six-month period? Of course, we can. Since we are now looking for order quantity, we modify our model to:

$$TC = \text{Total cost}$$

$$Co = \text{Ordering cost per order}$$

$$Cc = \text{Carrying cost}$$

$$Q = \text{Quantity ordered}$$

$$D = \text{Demand}$$

Our components for TC are total ordering costs and total carrying costs where

$$\text{Total Ordering Costs} = Co(D/Q)$$

And

$$\text{Total Carrying Cost} = Cc(Q/2)$$

So

$$TC = Co(D/Q) + Cc(Q/2)$$

$$dTC/dQ = -CoD/Q^2 + Cc/2$$

We set equal to 0 and solve for Q^*

$$Q^* = \sqrt{\frac{2C_o D}{C_c}} \qquad \text{(Eq. 1.8)}$$

In this EOQ model, we are finding the order quantity to minimize inventory costs such as holding costs, storage costs, and ordering costs.

First, for the six-month period we use:

$$d = 9,000$$

$$Co = \$80$$

$$Cc = \$0.10 * \$20 = \$2$$

$$EOQ \; Q^* =$$

$$\sqrt{\frac{2*9000*80}{2}} = 848.52 \approx 849 \text{ (round up)}$$

For the next six months,

$$D = 15,000$$

$$Co = \$80$$

$$Co = \$2 \text{ (as before)}$$

$$Q^* = 1095.445 \text{ or } 1096 \text{ (rounded up)}$$

The formula is independent of time.

We may also use calculus methods for facility location problems.

1.4 FACILITY LOCATION WITH AN OIL RIG LOCATION PROBLEM

This problem is adapted from Giordano et al. (2014) and considers an oil-drilling rig that is 8.5 miles offshore. The drilling rig is to be connected by an underwater pipe to a pumping station. The pumping station is connected by land-based pipe to a refinery, which is 14.7 miles down the shoreline from the drilling rig (see Figure 1.5). The underwater pipe costs \$31,575 per

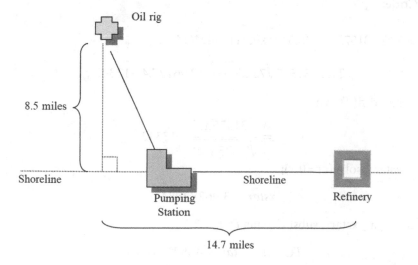

FIGURE 1.5   Schematics of Oil Rig Problem.

mile, and land-based pipe costs $13,342 per mile. You need to determine where to place the pumping station to minimize cost of the pipe.

Problem: Find a relationship between the location of the pumping station and the cost of the installation of the pipe to minimize costs.

Assumptions and variables: First, we assume no cost saving for the pipe if we purchase in larger lot sizes. We further assume no additional costs are incurred in preparing the terrain to lay the pipe.

1.4.1 Variables

x = The location of the pumping station along the horizontal distance from $x = 0$ to $x = 14.7$ miles.

TC = total cost of the pipe for both underwater and on shore piping. We develop the model as follows:

We use Pythagorean's Theorem for finding the underwater distance of the pipe that is the hypotenuse of the right triangle with height 8.5 miles and base = x. The hypotenuse is:

$\sqrt{8.5^2 + x^2}$. The length of the pipe on shore is 14.7 − x.

Total cost = $31,575 \sqrt{8.5^2 + x^2} + 13342 (14.7 - x)$.

Solving in R would provide the following solution:

R Code:

```
> TC:= 31575*sqrt(8.5^2+x^2)+13342*(14.7-x);
```

$$TC := 31575\sqrt{72.25 + x^2} + 196127.4 - 13342\,x$$

```
> cp:=diff(TC,x);
```

$$cp := \frac{31575\,x}{\sqrt{72.25 + x^2}} - 13342$$

```
> xstar:=solve(cp=0,x);
```

$$xstar := 3.962829844$$

```
> TC_at_xstar:=subs(x=xstar,TC);
```

$$TC\_at\_xstar := 439377.6878$$

```
> plot(TC,x=0.7, thickness=3, title='Total_Cost');
```

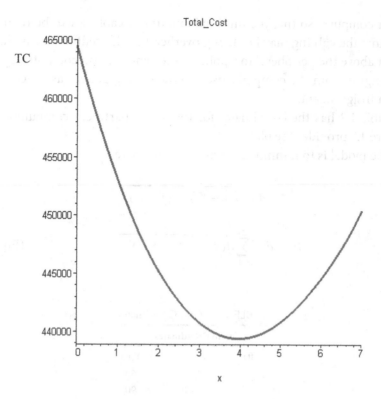

FIGURE 1.6   Plot of Total Cost.

Thus, if the pumping station is located at 14.7 − 3.963 = 10.737 miles from the refinery, we will minimize the total cost at a cost of $439,377.69. This is illustrated in Figure 1.6.

In Excel, we would use the Solver. In cell a1, it is type x, in cell a2, it is type 0. In cell a2, it is type objective function, and in cell b2, it is type = 31575*sqrt(8.5^2+(a2)^2)+13342*(14.7-(a2)). Highlight cell B2, open the Solver, click on min, enter cell a2 in changing cells, and under the options click on non-negativity and solve. This should return a minimum cost value of $439,377.69 for a grid location of (3.963, 0).

1.5 COMPUTER CABLING LOCATION OF CENTRAL COMPUTER

Consider a small company that is planning to install a central computer with cable links to five new departments (Fox et al., 2002). According to the floor plan, the peripheral computers to the five departments will be situated as shown in Figure 1.7. The company wishes to locate the central main computer so that the minimal amount of cabling will be required. Assume the cabling may be strung overhead in the ceiling panels from a point above the peripheral to a point to the central main computer. Ignore all lengths from the computer itself to the ceiling panels, as we consider them insignificant.

Table 1.2 has the coordinates for the five department computers, and Figure 1.7 provides the plot.

The model is to minimize the distance d, where

$$d = \sqrt{(x - X_i)^2 + (y - Y_i)^2}$$

$$d = \sum_{i=1}^{5} \sqrt{(x - X_i)^2 + (y - Y_i)^2} \qquad \text{(Eq. 1.9)}$$

TABLE 1.2   Grid Coordinates

Grid Coordinates	
X(ft)	Y(ft)
15	60
25	90
60	75
75	60
80	22

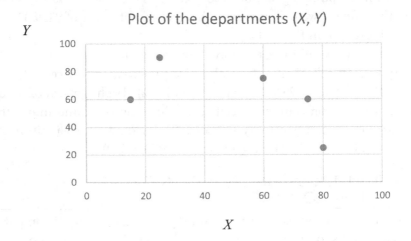

Plot of the departments (X, Y)

FIGURE 1.7   Plot of Grid Coordinates for the Computer Locations.

x	y		Guess	x	y
15	60	42.5909304		56.81842	68.07511
25	90	38.6408141			
60	75	7.62080023			
75	60	19.8941535			
80	25	48.9167738			
		157.663472	Min Z		

FIGURE 1.8   Screenshot from Excel.

We use the gradient search method (GRG nonlinear) on the Excel solve (Figure 1.8), and we find that $d = 157.663472$ ft with the central computer located at $(x^*, y^*) = (56.81842, 68.07511)$.

For a more complete description of this problem, please see Fox & Richardson (2002).

1.6 EXERCISES

1.1. Consider an industrial situation where it is necessary to set up an assembly line. Suppose that each time the line is set up, a cost c is incurred. Assume that c is in addition to the cost of producing any item and is independent of the amount produced. Suggest models for the production rate. Now assume a constant production rate k and a constant demand rate r. What assumptions are implied by the model in Figure 1.9? Assume a storage cost of s (in dollars per unit per day) and compute the optimal length of the production run P^* in order to minimize the total cost.

1.2. Use the EOQ model to obtain the realistic results for the lot size, q_0, the number of setups, S_0, and the total cost, C_t, for the following:

(a) $R = 100,00, C_0 = \$1,000, C_i = \$1$

(b) $R = 100,00, C_0 = \$10, C_i = \$1$

(c) $R = 100,00, C_0 = \$.50, C_i = \$1$

1.3. Consider an oil-drilling rig that is 10.5 miles offshore. The drilling rig is to be connected by underwater pipe to a pumping station. The pumping station is connected by a land-based pipe to a refinery, which is 15.7 miles down the shoreline from the drilling rig (see Figure 1.9). The underwater

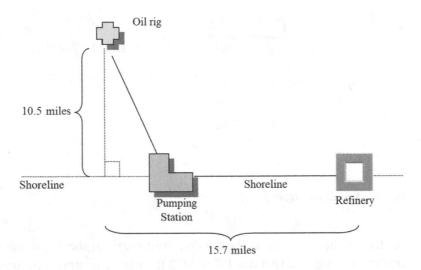

FIGURE 1.9   Schematics of Oil Rig Problem for Exercise 1.3.

TABLE 1.3   Location Coordinates
for Computers

Grid Coordinates	
X(ft)	Y(ft)
25	65
35	92
70	78
85	65
90	30

pipe costs $31,575 per mile, and land-based pipe costs $14,442 per mile. Determine where to place the pumping station to minimize cost of the pipe.

1.4. Consider a computer system that needs its main computer positioned to minimize the distance to the other computers. The grid coordinates of the other computers are in the following Table 1.3.

REFERENCES AND ADDITIONAL READINGS

Fox, William P. and Richardson, William H. (2002). Multivariable Variable Optimization When Calculus Fails: Gradient Search Methods in Nonlinear Optimization Using MAPLE, *Computers in Education Journal*, XII (2), pp. 2–11. September-December.

Giordano, F., Fox, W. and Horton, S. (2014). *A first course in mathematical modeling*, 5th Ed. Cengage Publishers, Belmont, CA.

CHAPTER 2

Product Mix

Linear Programming Problems

L INEAR PROGRAMMING IS A discipline focused on optimizing operations, usually with some constraints on the system. Typically, the
objective of linear programming is to maximize or minimize the value
based on some objective function of the problem. The basic assumption is that the problem is linear in nature and is subjected to the
constraints in the form of linear equations. Linear programming is
an important operations research technique that seeks to understand
the relationship between multiple variables and seeks to find the best
solution from various alternatives. Linear programming is widely used
in many fields such as economics, business, telecommunication, and
manufacturing.

We will start this chapter with a simple linear programming problem
discussion to help set the "table". You are the owner of a company that
manufactures two types of end tables: Oblong and rectangular. The rectangular table is by far the most popular and sells for $90 each, while the
oblong table sells for $70 each. The first internal argument that might come
up is "why should we manufacture the oblong table at all since they do
not sell as well as the rectangular table". This would be a simple decision
if there were limitless wood, man-hours for carpentry, man-hours for finishing, etc., then perhaps the argument would make sense. However, the
real world is constrained by labor, capital, and time (just to name a few
constraints). Good managers must take all these factors into consideration

DOI: 10.1201/9781003464969-2

before making decisions as these limitations affect the product mix, the cost of the operation, and the profitability of the business.

The decisions made regarding the number of each product we want to manufacture must consider the allocation of resources in a manner that will lead to either a minimization of cost or a maximization of profit. The two examples discussed in this chapter are representative of the type of problems a manager might face.

We present the concepts of the Simplex method using technology as typically you will use technology to solve and analyze such problems. One of the most important aspects of the Simplex method is sensitivity analysis. Sensitivity analysis provides additional information to managers about "what if" questions that might arise.

2.1 LINEAR PROGRAMMING PROBLEM INTRODUCTION

2.1.1 Formulations

A linear programming problem is a problem that requires an objective function to be maximized or minimized, subject to resource constraints. The key to formulating a linear programming problem is recognizing the *decision variables*. The objective function and all constraints are written in terms of these decision variables.

The two conditions necessary for a mathematical model to be a linear program (LP) are:

1. All continuous variables (i.e. can take fractional values).

2. A single objective (minimize or maximize).

   - The objective function and the constraints are linear, i.e., any term is either a constant or a constant multiplied by an unknown.

   - The decision variables must be nonnegative.

Linear programming problems are important, because many practical problems can be formulated as LPs, and there exists an algorithm (called the *simplex* algorithm) that enables us to solve LPs numerically relatively easily.

We will return later to the simplex algorithm for solving LPs, but for the moment, we will concentrate upon formulating LPs.

Some of the major application areas to which LP can be applied are:

- Blending

- Production planning

- Oil refinery management

- Distribution

- Financial and economic planning

- Manpower planning

- Blast furnace burdening

- Farm planning

We consider some specific examples of the types of problem that can be formulated as LPs. Note that here that the key to formulating LPs is *practice*. However, a useful hint is that common objectives for LPs are either to minimize *cost* or to *maximize profit*.

2.2 SIMPLE MANUFACTURING EXAMPLE

Consider the following problem statement: A company wants to can two new different drinks for the holiday season. It takes 2 hours to can one gross of Drink A, and it takes 1 hour to label the cans. It takes 3 hours to can one gross of Drink B, and it takes 4 hours to label the cans. The company makes $10 profit on one gross of Drink A and a $20 profit of one gross of Drink B. Given that we have 20 hours to devote to canning the drinks and 15 hours to devote to labeling cans per week, how many cans of each type of drink should the company package to maximize profits? Let's formulate the LP for this problem.

Problem: Maximize the profit of selling these new drinks.

Define decision variables:

X_1 = the number of gross cans produced for Drink A per week

X_2 = the number of gross cans produced for Drink B per week

Objective function:

$Z = 10X_1 + 20X_2$

Constraints:

(1) Canning with only 20 hours available per week:

$2X_1 + 3X_2 \leq 20$

(2) Labeling with only 15 hours available per week:

$X_1 + 4 X_2 \leq 15$

(3) Non-negativity restrictions:

$X_1 \geq 0$ (non-negativity of the production items)

$X_2 \geq 0$ (non-negativity of the production items)

Complete LP formulation:

$$\text{Maximize Z} = 10X_1 + 20X_2$$

Subject to:

$$2 X_1 + 3 X_2 \leq 20$$

$$X_1 + 4 X_2 \leq 15$$

$$X_1 \geq 0$$

$$X_2 \geq 0$$

2.3 FINANCIAL PLANNING

Now consider a simple banking example. A bank provides four kinds of loans to its personal customers, and these loans yield the following annual interest rates to the bank:

- First mortgage: 14%

- Second mortgage: 20%

- Home improvement: 20%

- Personal overdraft: 10%

The bank has a maximum foreseeable lending capability of $250 million and is further constrained by the policies:

1. First, mortgages must be at least 55% of all mortgages issued and at least 25% of all loans issued (in terms of $).

2. Second, mortgages cannot exceed 25% of all loans issued (in terms of $).

3. To avoid public displeasure and the introduction of a new windfall tax, the average interest rate on all loans must not exceed 15%.

Note here that these policy conditions, while potentially limiting the profit that the bank can make, also limit its exposure to risk in a particular area. It is a fundamental principle of risk reduction that risk is reduced by spreading money (appropriately) across different areas.

Formulate the bank's loan problem as an LP so as to maximize interest income while satisfying the policy limitations.

2.3.1 Financial Planning Formulation

Note here that as in *all* formulation exercises, we are translating a verbal description of the problem into an *equivalent* mathematical description.

A useful tip when formulating LPs is to express the variables, constraints, and objective in words before attempting to express them in mathematics.

2.3.2 Decision Variables

Essentially, we are interested in the amount (in dollars) the bank has loaned to customers in each of the four different areas (not in the actual number of such loans). Hence, let x_i = amount loaned in area i in millions of dollars (where $i = 1$ corresponds to first mortgages, $i = 2$ to second mortgages, etc.) and note that each $x_i \geq 0$ ($i = 1, 2, 3, 4$).

Note here that it is conventional in LPs to have all variables ≥ 0. Any variable (X, say) which can be positive *or* negative can be written as $X_1 - X_2$ (the difference of two new variables) where $X_1 \geq 0$ and $X_2 \geq 0$.

2.3.3 Constraints

(a) Limit on amount lent:

$$x_1 + x_2 + x_3 + x_4 \leq 250$$

(b) Policy condition 1:

$$x_1 \geq 0.55(x_1 + x_2)$$

i.e., first mortgages >= 0.55(total mortgage lending) and also

$$x_1 \geq 0.25(x_1 + x_2 + x_3 + x_4)$$

i.e., first mortgages $\geq$ 0.25(total loans)

(c) Policy condition 2:

$$x_2 \le 0.25(x_1 + x_2 + x_3 + x_4)$$

(d) Policy condition 3 – we know that the total annual interest is $0.14x_1 + 0.20x_2 + 0.20x_3 + 0.10x_4$ on total loans of $(x_1 + x_2 + x_3 + x_4)$. Hence, the constraint relating to policy condition (3) is:

$$0.14x_1 + 0.20x_2 + 0.20x_3 + 0.10x_4 \le 0.15(x_1 + x_2 + x_3 + x_4)$$

2.3.4 Objective Function

To maximize interest income (which is given above), i.e.,

$$\text{Maximize } Z = 0.14x_1 + 0.20x_2 + 0.20x_3 + 0.10x_4$$

2.4 BLENDING FORMULATION EXAMPLE

Consider the example of a manufacturer of animal feed who is producing feed mix for dairy cattle. In our example, the feed mix contains two active ingredients. One kg of feed mix must contain a minimum quantity of each of four nutrients as given below:

Nutrients: A B C D

Of grams: 90   50   20   2

The ingredients have the following nutrient values and cost:

	A	B	C	D	Cost/kg
Ingredient 1 (grams/kg)	10	80	40	10	40
Ingredient 2 (grams/kg)	200	150	20	0	0

What should be the amounts of active ingredients in 1 kg of feed mix that minimizes cost?

2.4.1 Blending Problem Formulation

2.4.1.1 Decision Variables

In order to solve this problem, it is best to think in terms of 1 kilogram of feed mix. That kilogram is made up of two parts – ingredient 1 and ingredient 2:

x_1 = amount (kg) of ingredient 1 in 1 kg of feed mix

x_2 = amount (kg) of ingredient 2 in 1 kg of feed mix

where $x_1 \geq 0, x_2 \geq 0$

Essentially, these variables (x_1 and x_2) can be thought of as the recipe telling us how to make up 1 kilogram of feed mix.

2.4.1.2 Constraints

- Nutrient constraints:

$100x_1 + 200x_2 \geq 90$ *(nutrient A)*

$80x_1 + 150x_2 \geq 50$ *(nutrient B)*

$40x_1 + 20x_2 \geq 20$ *(nutrient C)*

$10x_1 \geq 2$ *(nutrient D)*

- Balancing constraint (an *implicit* constraint due to the definition of the variables):

$x_1 + x_2 = 1$

2.4.1.3 Objective Function

Presumably to minimize cost, i.e.,

$$\text{Minimize } Z = 40x_1 + 60x_2$$

This gives us our complete LP model for the blending problem.

2.5 PRODUCTION PLANNING PROBLEM

A company manufactures four variants of the same table, and in the final part of the manufacturing process there are assembly, polishing, and packing operations. For each variant, the time required for these operations is shown in Table 2.1 (in minutes) as is the profit per unit sold.

Given the current state of the labor force, the company estimates that, each year, they have 100,000 minutes of assembly time, 50,000 minutes of polishing time, and 60,000 minutes of packing time available. How many of each variant should the company make per year and what is the associated profit?

TABLE 2.1   Table Assembly, Polish, Pack, and Profit

Variant	Assembly	Polish	Pack	Profit ($)
1	2	3	3	$1.50
2	4	2	3	$2.50
3	3	3	2	$3.00
4	7	4	5	$4.50

2.5.1 Decision Variables

Let x_i be the number of units of variant i ($i = 1, 2, 3, 4$) made per year where $x_i \geq 0$ $i = 1, 2, 3, 4$

2.5.2 Constraints

Resources for the operations of assembly, polishing, and packing:

$$2x_1 + 4x_2 + 3x_3 + 7x_4 \leq 100{,}000 \ (assembly)$$

$$3x_1 + 2x_2 + 3x_3 + 4x_4 \leq 50{,}000 \ (polishing)$$

$$2x_1 + 3x_2 + 2x_3 + 5x_4 \leq 60{,}000 \ (packing)$$

2.5.3 Objective Function

Maximize $Z = 1.5x_1 + 2.5x_2 + 3.0x_3 + 4.5x_4$

2.6 SHIPPING PROBLEM

Consider planning the shipment of needed items from the warehouses where they are manufactured and stored to the distribution centers where they are needed as shown in the introduction. There are three warehouses at different cities: Detroit, Pittsburgh, and Buffalo. They have 250, 130, and 235 tons of paper accordingly. There are four publishers in Boston, New York, Chicago, and Indianapolis. They ordered 75, 230, 240, and 70 tons of paper to publish new books.

There are the following costs in dollars of transportation of one ton of paper:

From\To	Boston (BS)	New York (NY)	Chicago (CH)	Indianapolis (IN)
Detroit (DT)	15	20	16	21
Pittsburgh (PT)	25	13	5	11
Buffalo (BF)	15	15	7	17

Management wants you to minimize the shipping costs while meeting the demand.

We define x_{ij} to be the travel from city i (1 is Detroit, 2 is Pittsburg, 3 is Buffalo) to city j (1 is Boston, 2 is New York, 3 is Chicago, and 4 is Indianapolis).

Objective function is:

$$Minimize\ Z = 15x_{11} + 20x_{12} + 16x_{13} + 21x_{14} + 25x_{21} + 13x_{22}$$
$$+ 5x_{23} + 11x_{24} + 15x_{31} + 15x_{32} + 7x_{33} + 17x_{34}$$

Subject to (constraints):

$$x_{11} + x_{12} + x_{13} + x_{14} \leq 250\ (availability\ in\ Detroit)$$
$$x_{21} + x_{22} + x_{23} + x_{24} \leq 130\ (availability\ in\ Pittsburg)$$
$$x_{31} + x_{32} + x_{33} + x_{34} \leq 235\ (availability\ in\ Buffalo)$$
$$x_{11} + x_{21} + x_{31} \geq 75\ (demand\ Boston)$$
$$x_{12} + x_{22} + x_{32} \geq 230\ (demand\ New\ York)$$
$$x_{13} + x_{23} + x_{334} \geq 240\ (demand\ Chicago)$$
$$x_{14} + x_{24} + x_{34} \geq 70\ (demand\ Indianapolis)$$
$$x_{ij} \geq 0$$

2.7 PRODUCT MIX

Many applications in business and economics involve a process called optimization. In optimization problems, you are asked to find the minimum or the maximum result. This section illustrates the strategy in graphical simplex of linear programming. We will restrict ourselves in this graphical context to two dimensions. Variables in the Simplex method are restricted to positive variables (e.g., $x \geq 0$).

2.7.1 Memory Chips for CPUs

Let's start with a manufacturing example. Suppose that a small business wants to know how many of two types of high-speed computer chips to manufacture weekly to maximize their profits. First, we need to define our *decision variables*.

Let,

x_1 = number of high-speed chip type A to produce weekly

x_2 = number of high-speed chip type B to produce weekly

TABLE 2.2 Chip Manufacturing Information

	Chip A	Chip B	Quantity Available
Assembly time (hours)	2	4	1,400
Installation time (hours)	4	3	1,500
Profit (per unit)	140	120	

The company reports a profit of $140 for each type A chip and $120 for each type B chip sold. The production line reports the following information (Table 2.2):

The constraint information from the table becomes inequalities that are written mathematically as:

Constraints are:

$$2x_1 + 4x_2 \leq 1400 \text{ (assembly time)}$$

$$4x_1 + 3x_2 \leq 1500 \text{ (installation time)}$$

$$x_1 \geq 0, x_2 \geq 0$$

The profit equation is:

Objective function:

$$\text{Profit } Z = 140x_1 + 120x_2$$

The LP formulation is:

$$\text{Profit } Z = 140x_1 + 120x_2$$

Subject to:

$$2x_1 + 4x_2 \leq 1400 \text{ (assembly time)}$$

$$4x_1 + 3x_2 \leq 1500 \text{ (installation time)}$$

$$x_1 \geq 0, x_2 \geq 0$$

Solution with technology in Excel involves using the SimplexLP of the Solver (Figure 2.1).

The Solver report is given in Figure 2.2.

The sensitivity report from Excel is in Figure 2.3.

Decision Variables		(1) Initial as 0	
x1	180		
x2	260		
Objective functions			
Z =	56400		
Constraints	Used	Resource capacity	
Assembly	1400	1400	
Installation	1500	1500	

FIGURE 2.1   Product Mix LP Setup in Excel.

Microsoft Excel 12.0 Answer Report
Worksheet: [Book3]Sheet1
Report Created: 12/7/2020 10:10:46 AM

Target Cell (Max)

Cell	Name	Original Value	Final Value
$C$10	Z=	0	56400

Adjustable Cells

Cell	Name	Original Value	Final Value
$C$5	x1	0	180
$C$6	x2	0	260

Constraints

Cell	Name	Cell Value	Formula	Status	Slack
$C$14	Assembly Used	1400	$C$14<=$D$14	Binding	0
$C$15	Installation Used	1500	$C$15<=$D$15	Binding	0
$C$5	x1	180	$C$5>=0	Not Binding	180
$C$6	x2	260	$C$6>=0	Not Binding	260

FIGURE 2.2   Product Mix LP Solution from Excel.

Microsoft Excel 12.0 Sensitivity Report
Worksheet: [Book3]Sheet1
Report Created: 12/7/2020 10:10:46 AM

Adjustable Cells

Cell	Name	Final Value	Reduced Cost	Objective Coefficient	Allowable Increase	Allowable Decrease
$C$5	x1	180	0	140	20	80
$C$6	x2	260	0	120	160	15

Constraints

Cell	Name	Final Value	Shadow Price	Constraint R.H. Side	Allowable Increase	Allowable Decrease
$C$14	Assembly Used	1400	6	1400	600	650
$C$15	Installation Used	1500	32	1500	1300	450

FIGURE 2.3   Product Mix Sensitivity Report from Excel.

The solution in this case is to manufacture 180 of x_1 and 260 of x_2 to yield a profit of $56,400. From the sensitivity analysis, we find that we have two shadow prices: One for assembly and one for installation. The shadow prices are $\lambda_1 = 6$ and $\lambda_2 = 32$. This implies that if we increase the hours of the resource for assembly from 1,400 to 1,401, our profit increases by approximately $6, and if instillation increases from 1,500 to 1,501, that profit increases by approximately $32. If the cost of an additional hour is the same and we can only increase one resource, instillation is the most profitable resource to increase.

2.8 SUPPLY CHAIN OPERATIONS (GASOLINE DISTRIBUTION)

Here, we present a linear programming model for supply chain design. We consider producing a new mixture of gasoline. We desire to maximize profit for distribution and sales of the new mixture. There is a supply chain involved with a product that must be modeled. The product is made up of components that are produced separately (Table 2.3).

Blending information is given in Table 2.4.

Each barrel of blend 1 can be sold for $1.10, and each barrel of blend 2 can sell for $1.20. Long-term contracts require at least 10,000 barrels of each blend to be produced.

TABLE 2.3 Oil Supply Chain, Oil Availability, and Cost

Crude Oil Type	Cost/ Barrel	Number of Available Barrels (000's of Barrels)
X_{10}	$0.30	6,000
X_{20}	$0.40	10,000
X_{30}	0.48	12,000

TABLE 2.4 Supply-Oil-Blending Requirements

Gasoline	Compound X_{10} (%)	Compound X_{20} (%)	Compound X_{30} (%)	Expected Demand (000 of Barrels)
Blend 1	> 0.30 X_{10}	<.50 X_{20}	> 0.3 X_{30}	*14,000*
Blend 2	< 4 B	> 0.35 X_{20}	< 0.4 X_{30}	*22,000*

Let i = Crude type 1, 2, 3 (X_{10}, X_{20}, X_{30}, respectively)

Let j = Gasoline blend type 1, 2 (Blend 1, Blend 2)

We define the following decision variables:

Gij = Amount of crude i used to produce gasoline j

For example, G_{11} = Amount of crude X_{10} used to produce Blend 1.

G_{12} = Amount of crude type X_{20} used to produce Blend 1
G_{13} = Amount of crude type X_{30} used to produce Blend 1
G_{12} = Amount of crude type X_{10} used to produce Blend 2
G_{22} = Amount of crude type X_{20} used to produce Blend 2
G_{32} = Amount of crude type X_{30} used to produce Blend 2

Revenue = 1.1 × (G11+ G12 + G13) + 1.2(G21 + G22 + G23)

Cost = .3 (G11 + G21) + .4 (G12 + G22) + 0.48 (G13 + G23)

LP formulation is:

Maximize Profit = Revenue − Cost

Subject to:

Oil availability:

$G11 + G21 \leq 6{,}000$

$G12 + G22 \leq 10{,}000$

$G13 + G23 \leq 12{,}000$

Contracts:

$G11 + G12 + G13 \geq 10{,}000$

$G21 + G22 + G23 \geq 10{,}000$

Composition or product mix in mixture format:

$G11/(G11 + G12 + G13) \geq 0.30$

$G12/(G11 + G12 + G13) \leq 0.50$

$G13/(G11 + G12 + G13) \geq 0.30$

$G21/(G21 + G22 + G23) \leq 0.4$

$G22/(G21 + G22 + G23) \geq 0.35$

$G23/(G21 + G22 + G23) \leq 0.40$

All decision variables are nonnegative.

Gasoline Distribution Case Study with Excel is given in Figure 2.4.

We examine the sensitivity report (Figure 2.5).

In our analysis of the sensitivity report, we noticed a possible alternate optimal solution. We obtained the second solution. Two solutions are found yielding a maximum profit of $21,040.00 (Figure 2.5 and Table 2.5).

Any oil available at a cost less than $0.90/barrel would make an additional profit. Also, if we could relax the long-term contract from 10,000 barrels to less for Blend 1 at a cost less than $0.72, then we could make an additional profit. The shadow price is an important concept. It measures the increase or decrease of the objective function due to a unit increase in an available resource.

2.9 PRODUCT MIX WITH LINDO

The following example provides a product mix problem but uses LINDO to solve the problem.

G11	3000			
G12	2200			
G13	4800			
G21	3000			
G22	7800			
G23	7200			
OBJF				
	21040			
CONSTRAINTS				
	USED	RHS		
	6000	6000		
	10000	10000		
	12000	12000		
	10000	10000		
	18000	10000		
	0	0	>	
	-2800	0	<	
	1800	0	>	
	-4200	0	<	
	1500	0	>	
	0	0	<	

FIGURE 2.4   Oil Supply Solver Setup in Excel.

Given the production requirements listed in the LP formulation, we can use LINDO to find a solution.

```
LP OPTIMUM FOUND AT STEP 2:

     OBJECTIVE FUNCTION VALUE:

   1)     56,400.00

VARIABLE VALUE REDUCED COST

     X₁    180.000.00
     X₂    260.000.00

ROW SLACK OR SURPLUS DUAL PRICES

   2)    0.00       6.00
   3)    0.00      32.00

NO. OF ITERATIONS = 2
```

Variable Cells

Cell	Name	Final Value	Reduced Cost	Objective Coefficient	Allowable Increase	Allowable Decrease
$C$5	G11	3000	0	0.8	0	1E+30
$C$6	G12	2200	0	0.7	0.333333333	0
$C$7	G13	4800	0	0.62	0	1.2
$C$8	G21	3000	0	0.9	1E+30	0
$C$9	G22	7800	0	0.8	0	0.333333333
$C$10	G23	7200	0	0.72	1E+30	0

Constraints

Cell	Name	Final Value	Shadow Price	Constraint R.H. Side	Allowable Increase	Allowable Decrease
$C$18	USED	6000	0.9	6000	2000	3000
$C$19	USED	10000	0.8	10000	4500	5500
$C$20	USED	12000	0.72	12000	3666.666667	3000
$C$21	USED	10000	-0.1	10000	8000	7333.333333
$C$22	USED	18000	0	10000	8000	1E+30
$C$23	USED	0	0	0	2200	1500
$C$24	USED	-2800	0	0	1E+30	2800
$C$25	USED	1800	0	0	1800	1E+30
$C$26	USED	-4200	0	0	1E+30	4200
$C$27	USED	1500	0	0	1500	1E+30
$C$28	USED	0	0	0	1500	2200

FIGURE 2.5 Oil Supply Sensitivity Analysis from Excel.

TABLE 2.5 Oil Supply Solution

Decision Variable	Z = $21040.00	Z = $21,040.00
G_{11}	3,000	3,414.41
G_{12}	2,200	1,483.99
G_{13}	4,800	5,101.60
G_{21}	3,000	2,585.59
G_{22}	7,800	8,516.01
G_{23}	7,200	6,898.40

RANGES IN WHICH THE BASIS IS UNCHANGED: OBJ COEFFICIENT RANGES

Variable	Current coef	Allowable increase	Allowable decrease
X_1	140.00	20.00	80.00
X_2	120.00	160.00	15.00

RIGHT-HAND SIDE RANGES

Row	Current rhs	Allowable increase	Allowable decrease
2	1,400	600.00	650.00
3	1,500	1,300	450.00

Row (Basis)	(Basis)	X1	X2	Slk2	Slk3
ART	0.00	0.00	6.00	32.00	56,400.00
X_1	0.00	1.00	0.40	0.20	260.00
X_2	1.00	0.00	-0.30	0.40	180.00

LINDO provides the solution and range sensitivity analysis (Table 2.6).

2.10 EXERCISES

2.1. With the rising cost of gasoline and increasing prices to consumers, the use of additives to enhance performance of gasoline is being considered. Consider two additives, Additive 1 and Additive 2. The following conditions must hold for the use of additives:

TABLE 2.6 LINDO Tableau for the Product Mix Problem.

Row (Basis)	(Basis)	X1	X2	SLK2	SLK3
ART	−140.00	−120.00	0.00	0.00	0.00
SLK2	2.00	4.00	1.00	0.00	1,400
SLK3	4.00	3.00	0.00	1.00	1,500
ART	−140.00	−120.00	0.00	0.00	0.00

- Harmful carburetor deposits must not exceed 1/2 lb per car's gasoline tank.

- The quantity of Additive 2 plus twice the quantity of Additive 1 must be at least 1/2 lb per car's gasoline tank.

- 1 lb of Additive 1 will add 10 octane units per tank, and 1 lb of Additive 2 will add 20 octane units per tank. The total number of octane units added must not be less than six.

- Additives are expensive and cost $1.53/lb for Additive 1 and $4.00/lb for Additive 2.

We want to determine the quantity of each additive that will meet the above restrictions and will minimize their cost. After we solve, assume now that the manufacturer of additives has the opportunity to sell you a nice TV special deal to deliver at least 0.5 lb of Additive 1 and at least 0.3 lb of Additive 2. Use graphical LP methods to help recommend whether you should buy this TV offer. Support your recommendation.

2.2. A farmer has 30 acres on which to grow tomatoes and corn. Each 100 bushels of tomatoes require 1,000 gallons of water and 5 acres of land. Each 100 bushels of corn require 6,000 gallons of water and 2.5 acres of land. Labor costs are $1 per bushel for both corn and tomatoes. The farmer has 30,000 gallons of water and $750 in capital available. He knows that he cannot sell more than 500 bushels of tomatoes or 475 bushels of corn. He estimates a profit of $2 on each bushel of tomatoes and $3 of each bushel of corn. How many bushels of each should he raise to maximize profits?

After the initial solution, consider the following modifications: Assume now that farmer can sign a nice contract with a grocery store to grow and deliver at least 300 bushels of tomatoes and at least 500 bushels of corn. Use graphical LP methods to help recommend a decision to the farmer. Support your recommendation.

If the farmer can obtain an additional 10,000 gallons of water for a total cost of $50, is it worth it to obtain the additional water? Determine the new optimal solution caused by adding this level of resource.

2.3. *Fire Stone Tires* headquartered in Akron, Ohio, has a plant in Florence, South Carolina, which manufactures two types of tires: SUV 225 Radials and SUV 205 Radials. Demand is high because of the recent recall of tires. Each 100 SUV 225 Radial requires 100 gallons of synthetic plastic and 5 lbs of rubber. Each 100 SUV 205 Radial requires 60 gallons synthetic plastic and 2.5 lbs of rubber. Labor costs are \$1 per tire for each type tire. The manufacturer has weekly quantities available of 660 gallons of synthetic plastic, \$750 in capital, and 300 lbs of rubber. The company estimates a profit of \$3 on each SUV 225 Radial and \$2 on each SUV 205 Radial. How many of each type tire should the company manufacture in order to maximize their profits? After solving the initial problems, consider the following modifications:

Assume now that the manufacturer can sign a nice contract with a tire outlet store to deliver at least 500 SUV 225 Radial tires and at least 300 SUV 205 radial tires. Use graphical LP methods to help recommend a decision to the manufacturer.

If the manufacturer can obtain an additional 1,000 gallons of synthetic plastic for a total cost of \$50, is it worth it to obtain this amount? Determine the new optimal solution caused by adding this level of resource.

If the manufacturer can obtain an additional 20 lbs of rubber for \$50, should they obtain the rubber? Determine the new solution caused by adding this amount.

2.4. Consider a toy maker company that carves wooden soldiers. The company specializes in two types: Confederate soldiers and union soldiers. The estimated profit for each is \$28 and \$30, respectively. A Confederate soldier requires 2 units of lumber, 4 hours of carpentry, and 2 hours of finishing to complete. A Union soldier requires 3 units of lumber, 3.5 hours of carpentry, and 3 hours of finishing to complete. Each week the company has 100 units of lumber delivered. The workers can provide at most 120 hours of carpentry and 90 hours of finishing. Determine the number of each type of wooden soldiers to produce to maximize the weekly profits.

2.5. A company wants to bottle two different drinks for the holiday season. It takes 3 hours to bottle one gross of Drink A, and it takes 2 hours to label the bottles. It takes 2.5 hours to bottle one gross of Drink B, and it takes 2.5 hours to label the bottles. The company makes \$15 profit on one gross of Drink A and an \$18 profit of one gross of Drink B. Given that we have 40 hours to devote to bottling the drinks and 35 hours to devote to labeling bottles per week, how many bottles of each type of drink should the company package to maximize profits?

2.6. The Seafarer Toy Company wishes to make three models of ships to maximize their profits. They found that a model steamship takes 1 hour of work of the cutter, 2 hours of work of the painter, and 4 hours of work of the assembler; it produces a profit of \$6.00. The sailboat takes 3 hours of work of the cutter, 3 hours of work of the painter, and 2 hours of the work of the assembler. It produces a $3.00 profit. For making the submarine, it takes the cutter 1 hour, the painter 3 hours, and the assembler 1 hour. It produces a profit of $2.00. The cutter is only available for 45 hours per week, the painter for 50 hours, and the assembler for 60 hours. Assuming that they sell all the ships that they make, formulate this LP to determine how many ships of each type that Seafarer should produce.

2.7. To produce 1,000 tons of nonoxidizing steel for BMW engine valves, at least the following units of manganese, chromium, and molybdenum will be needed weekly: 10 units of manganese, 12 units of chromium, and 14 units of molybdenum (1 unit is 10 lbs). These materials are obtained from a dealer who markets these metals in three sizes – small (S), medium (M), and large (L). One S case costs $9 and contains 2 units of manganese, 2 units of chromium, and 1 unit of molybdenum. One M case costs $12 and contains 2 units of manganese, 3 units of chromium, and 1 unit of molybdenum. One L case costs $15 and contains 1 unit of manganese, 1 unit of chromium, and 5 units of molybdenum. How many cases of each kind (S, M, L) should be purchased weekly so that we have enough manganese, chromium, and molybdenum at the smallest cost?

REFERENCES AND ADDITIONAL READINGS

Fox, W. P. (2017). *Mathematical Modeling for Business Analytics*. Taylor and Francis, CRC Press. December.

Fox, W. P. (2021). *Nonlinear Optimization*. Taylor and Francis, CRC Press.

Fox, W. P. and Bauldry, W. (2019). *Problem Solving with Maple*. Volume 1. Taylor and Francis, CRC Press.

Fox, W. P. and Bauldry, W. (2020). *Problem Solving with Maple*. Volume II. Taylor and Francis, CRC Press.

Fox, W. P. and Burks, R. E. (2019a). *Advanced Mathematical Modeling with Technology*. Taylor and Francis, CRC Press.

Fox, W. P. and Burks, R. E. (2019b). *Applications of Operations Research and Management Science for Military Decision Making*. Springer Series.

CHAPTER 3

Transportation and Shipping Problems

ALTHOUGH WE MENTIONED SHIPPING in Chapter 1, we devoted this chapter to strictly transportation and shipping problems.

3.1 TRANSPORTATION AND SHIPPING WAREHOUSE PROBLEM

We will start this chapter with a simple example of a transportation and shipping problem.

Three suppliers A, B, and C each produce an item that they need to deliver to companies W, X, Y, and Z. The stock held at each supplier and the demand from each company is known in advance. The cost, in dollars, of transporting one load of the item from the supplier to the company is also known. This information, which is needed to model this situation, is provided in Table 3.1.

As seen in this cost/matrix table, the amount of supply equals the amount of demand. If this is not the case where supply does not equal demand, we simply introduce a dummy variable to absorb the excess supply with transportation cost all equal to zero. We will explore this point later in the chapter. We also point out that since all the values in the table are integers, the solutions to all unknowns also will be integer solutions (see Theorem 7.3.1 from Albright and Fox).

Our solution goal is to minimize total costs while meeting supply and demand goals. We saw in Chapter 2 how to formulate a linear program.

DOI: 10.1201/9781003464969-3

TABLE 3.1   Cost/Matrix Table for Supply/Demand

Company/ Supplier	W	X	Y	Z	Stock Available
A	180	110	130	290	14
B	190	250	150	280	16
C	240	270	190	120	20
Demand	11	15	14	10	50

3.1.1 Decision Variables

x_{ij} = From suppler i to company demand j for i = 1, 2, 3 and j = 1, 2, 3, 4.

The model formulation yields:

Objective function: Minimize

$$C = 180\, x_{11} + 110\, x_{12} + 130\, x_{13} + 290\, x_{14} + 190\, x_{21} + 250\, x_{22} + 150\, x_{23}$$
$$+ 280\, x_{24} + 240\, x_{31} + 270\, x_{32} + 190\, x_{33} + 120\, x_{3}$$

Subject to:

$$x_{11} + x_{12} + x_{13} + x_{14} = 14$$
$$x_{21} + x_{22} + x_{23} + x_{24} = 16$$
$$x_{31} + x_{32} + x_{33} + x_{34} = 20$$
$$x_{11} + x_{21} + x_{31} + x_{41} = 11$$
$$x_{12} + x_{22} + x_{32} + x_{42} = 15$$
$$x_{13} + x_{23} + x_{33} + x_{43} = 14$$
$$x_{14} + x_{24} + x_{34} + x_{44} = 10$$
$$x_{ij} \geq 0 \text{ for all } i = 1, 2, 3 \text{ and } j = 1, 2, 3, 4$$

3.1.2 Solution

Because of the theorem that we referenced earlier, we may solve this as a linear programming problem using any technologies, such as Excel or LINDO, and we obtain our integer solution. We obtain a total cost of $7,560 when x_{12} = 4, x_{21} = 11, x_{23} = 5, x_{32} = 1, x_{32} = 9, x_{34} = 10. This allows supply to equal demand. We used Excel to obtain our results in Table 3.2.

3.1.3 Modification to the Warehouse Problem

Let's modify our previous problem just slightly so that supply does not equal demand. We will allocate more supply than demand needed. We display the information in Table 3.4.

TABLE 3.2 Excel Linear Program Solution

	W	X	Y	Z	RHS
A	0	14	0	0	14
B	11	0	5	0	16
C	0	1	9	10	20
	11	15	14	10	

Total Cost = $7,560

TABLE 3.3 Supplier/Demand Shipping Route

	W	X	Y	Z	RHS
A	0	0	4	10	14
B	0	15	1	0	16
C	11	0	9	10	20
	11	15	14	10	

Total Cost = $11,670

TABLE 3.4 Cost Matrix for Modified Warehouse Problem

	Company W	Company X	Company Y	Company Z	Stock Available (Loads)
Supplier A	180	110	130	290	21
Supplier B	190	250	150	280	19
Supplier C	240	270	190	120	22
Demand (loads)	13	10	12	20	55/62

Our formulation is similarly done as in the Warehouse Problem (3.2).

Minimize:

$$C = 180\, x_{11} + 110\, x_{12} + 130\, x_{13} + 290\, x_{14} + 190\, x_{21} + 250\, x_{22} + 150\, x_{23}$$
$$+ 280\, x_{24} + 240\, x_{31} + 270\, x_{32} + 190\, x_{33} + 120\, x_{34} + 0\, x_{15} + 0x_{25} + 0x_{35}.$$

Subject to:

$$x_{11} + x_{12} + x_{13} + x_{14} = 21$$
$$x_{21} + x_{22} + x_{23} + x_{24} = 19$$
$$x_{31} + x_{32} + x_{33} + x_{34} = 22$$
$$x_{11} + x_{21} + x_{31} + x_{41} = 13$$
$$x_{12} + x_{22} + x_{32} + x_{42} = 10$$

$$x_{13} + x_{23} + x_{33} + x_{43} = 12$$
$$x_{14} + x_{24} + x_{34} + x_{44} = 20$$
$$x_{ij} \geq 0 \text{ for all } I = 1, 2, 3 \text{ and } j = 1, 2, 3, 4, 5 \text{ (dummy variable)}$$

3.1.3.1 Solution

Our solution methodology requires the use of a dummy variable that has no transportation cost but has demand equal to the excess supply (in this case seven items). Again, since all our inputs are integers, the Simplex procedure will solve it and yields integer solutions.

We may solve this as a linear programming using any technology, such as Excel or LINDO, to obtain our integer solution. We obtain a total cost of $7,550 when $x_{12} = 10$, $x_{13} = 11$, $x_{21} = 13$, $x_{23} = 1$, $x_{34} = 20$, and our dummy variables are $x_{25} = 5$, $x_{35} = 2$, see Table 3.5.

A shipper having m warehouses with supply a_i of goods at his ith warehouse must ship goods to n geographically dispersed retail centers, each with a given customer demand e_j, which must be met. The objective is to determine the minimum possible transportation costs, given that the unit cost of transportation between the ith warehouse and the jth retail center is C_{ij}.

In the following problem, the goal is to find the most effective way to transport the goods. The supply at each source is designated, and the demand at each destination is also given. For example, source 3 has 800 units available, and destination 1 needs at least 1,100 units. Each route from one source to one destination is assigned a unit transportation cost.

3.2 TRANSPORTATION NETWORK

Consider the following problem posed in networks form (Figure 3.1):

We have S1 supplying D1 with 35, D2 with 30, D3 with 40, and D4 with 32 units.

We have S2 supplying D1 with 37, D2 with 40, D3 with 42, and D4 with 25 units.

TABLE 3.5   Modified Warehouse Problem Solution

	W	X	Y	Z	Dummy	RHS
A	0	10	11	0	0	21
B	13	0	1	0	5	19
C	0	0	0	20	2	22
	13	10	12	20	7	

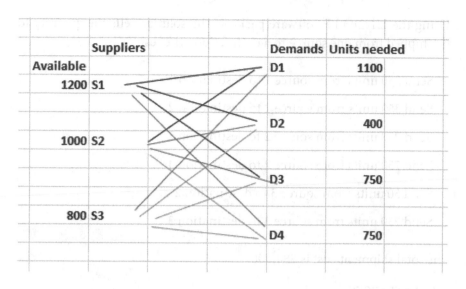

FIGURE 3.1 Transportation Network Problem Flow Diagram.

We have S3 supplying D1 with 40, D2 with 15, D3 with 10, and D4 with 28 units.

Decision variable:

Let x_{ij} = Amount send from supplier i (i = 1, 2, 3) to demand j (j = 1, 2, 3, 4)

Objective function:

$$Z = 35\, x_{11} + 30\, x_{12} + 40\, x_{13} + 32\, x_{14} + 37\, x_{21}, 40\, x_{22} + 42\, x_{23}$$
$$+ 25\, x_{25} + 40\, x_{31} + 15 x_{32} + 10\, x_{33} + 28\, x_{34}$$

Subject to:

$$x_{11} + x_{12} + x_{13} + x_{14} \leq 1200$$
$$x_{21} + x_{22} + x_{23} + x_{24} \leq 1100$$
$$x_{31} + x_{32} + x_{33} + x_{34} \leq 800$$
$$x_{11} + x_{21} + x_{31} \geq 1100$$
$$x_{12} + x_{22} + x_{32} \leq 400$$
$$x_{13} + x_{23} + x_{33} \geq 750$$
$$x_{14} + x_{24} + x_{34} \geq 750$$
$$x_{ij} \geq 0$$

Using the LINDO LP software package, the solution tells the quantity to be shipped from one source to a destination. The results are:

Send 850 units from source 1 to destination 1

Send 350 units from source 1 to destination 2

Send 250 units from source 2 to destination 1

Send 750 units from source 2 to destination 4

Send 50 units from source 3 to destination 2

Send 750 units from source 3 to destination 3

The total shipment cost is $84,000.

3.3 EXERCISES

3.1. Given we want to ship from companies A, B, C to Suppliers W, X, Y, Z as shown in Table 3.6. Determine the optimal solution.

3.2. Given we want to ship from companies A, B, C to Suppliers W, X, Y, Z as shown in Table 3.7. Determine the optimal solution.

TABLE 3.6 Transportation Shipping Information for Exercise 3.1

	Company/Supplier				
	W	X	Y	Z	Available
A	280	210	230	290	28
B	210	300	175	280	32
C	240	250	195	130	40
Demand	22	30	28	20	

TABLE 3.7 Transportation Shipping Information for Exercise 3.2

	Company/Supplier				
	W	X	Y	Z	Available
A	200	175	250	200	28
B	230	300	175	245	32
C	250	230	185	130	40
Demand	22	30	28	20	

TABLE 3.8   Transportation Shipping Costs for Exercise 3.3

	Plants/Sites			
	A	B	C	Supply
1	4	3	8	300
2	7	5	9	300
3	4	5	5	100
Demand	200	200	300	

3.3. A concrete company supplied concrete from three plants {1, 2, 3} to three works sites {A, B, C}. The plants can supply the following number of tons per week – Plant 1: 300, Plant 2: 300, and Plant 3: 100. The requirements of the sites, in tons per week, are Site A: 200, Site B: 200, and Site C: 300. Determine the optimal transportation solution when the costs associated in hundreds of dollars are as indicated in Table 3.8.

REFERENCES AND ADDITIONAL READINGS

Albright, B., Fox, W.P. (2020). *Mathematical Modeling with Excel*. 2nd Ed. Taylor and Francis, CRC Press.

Burks, R.E. (2006). "An Adaptive Tabu Search Heuristic for the Location Routing Pickup and Delivery Problem with Time Windows with a Theater Distribution Application," Ph.D. Dissertation, Air Force Institute of Technology.

Burks, R.E., Morre, J., Barnes, B., Bell, J. (2010). "Solving the Theater Distribution Problem with Tabu Search," *Military Operations Research*, 15:4, 5–26.

Assignment Models

TYPICALLY, WE HAVE A group of n "applicants" applying for n "jobs", and the nonnegative cost C_{ij} of assigning the ith applicant to jth job is known. The objective is to assign one job to each applicant in such a way as to achieve the minimum possible total cost. Define binary variables X_{ij} with value of either 0 or 1. When $X_{ij} = 1$, it indicates that we should assign applicant i to job j. Otherwise ($X_{ij} = 0$), we should not assign applicant i to job j.

A special case of the transportation problem is just discussed in Chapter 3 is the assignment problem, which occurs when each supply is 1 and each demand is 1. In this case, the integrality implies that every supplier will be assigned one destination, and every destination will have one supplier. The costs give the basis for assigning a supplier and destination to each other.

Suppose we want to impose the condition that either person i should not perform job j or person k should not perform job m. That is $X_{ij}.X_{km} = 0$. This nonlinear condition is equivalent to the linear constraint $X_{ij} + X_{km} \leq 1$. This constraint should be added to the set of constraints as a side constraint. With this additional constraint, the AP becomes a binary ILP, which could be solved by many software packages such as Excel or LINDO.

Let's proceed.

4.1 TRAINING CENTERS AND OFFICES

In the following problem, the goal is to assign people to particular tasks while minimizing the total cost. The objective function considers the cost involved for each person to do a particular task. The constraints say that each person must be assigned a task, and each task must be given to a person.

DOI: 10.1201/9781003464969-4

TABLE 4.1   Distance between Training Centers

	A	B	C	D	E
I1	14	7	3	7	27
I2	20	7	12	6	30
I3	10	3	4	5	21
I4	8	12	7	12	21
I5	13	25	24	26	8

A major company has training centers in five cities (let's call them {A, B, C, D, E}). As the manager, you must decide how to assign training teams from five offices to the training centers ({Let's call the offices I1, I2, I3, I4, I5}). What we know is the distance between each (in hundreds) of miles as indicated in Table 4.1. This table has the same format as the transportation problems covered in Chapter 3.

Let x_{ij} be the assignment of training team i to training center j. To save costs, we desire to minimize miles. We define c_{ij} as the cost (usually given in dollars or miles).

The above definition can be developed into a mathematical model as follows:

Determine $x_{ij} > 0$ $(i, j = 1, 2, 3, \ldots n)$ in order to

Minimize

$$\sum_{j=1}^{n} \sum_{i=1}^{n} c_{ij} x_{ij}$$

Subject to the constraints:

$$\sum_{i=1}^{n} x_{ij} \text{ for } j = 1, 2, \ldots, n$$

$$\sum_{j=1}^{n} x_{ij} \text{ for } i = 1, 2, \ldots, n$$

and x_{ij} is either zero or one.

Objective function: Minimize

$C = 14\,x_{11} + 7\,x_{12} + 3\,x_{13} + 7\,x_{14} + 27\,x_{15} + 20\,x_{21} + 7\,x_{22} + 12\,x_{23} + 6\,x_{24}$
$\qquad + 30\,x_{25} + 10\,x_{31} + 3\,x_{32} + 4\,x_{33} + 5\,x_{34} + 21\,x_{35} + 8\,x_{41} + 12\,x_{42} + 7\,x_{43}$
$\qquad + 12\,x_{44} + 21\,x_{45} + 13\,x_{51} + 25\,x_{52} + 24\,x_{53} + 26\,x_{54} + 8\,x_{25}$

Subject to:

$$x_{11} + x_{12} + x_{13} + x_{14} + x_{15} = 1$$
$$x_{21} + x_{22} + x_{23} + x_{24} + x_{25} = 1$$
$$x_{31} + x_{32} + x_{33} + x_{34} + x_{35} = 1$$
$$x_{41} + x_{42} + x_{43} + x_{44} + x_{45} = 1$$
$$x_{51} + x_{52} + x_{53} + x_{54} + x_{55} = 1$$
$$x_{11} + x_{21} + x_{31} + x_{41} + x_{51} = 1$$
$$x_{12} + x_{22} + x_{32} + x_{42} + x_{52} = 1$$
$$x_{13} + x_{23} + x_{33} + x_{43} + x_{53} = 1$$
$$x_{14} + x_{24} + x_{34} + x_{44} + x_{54} = 1$$
$$x_{15} + x_{25} + x_{35} + x_{45} + x_{55} = 1$$
$$x_{ij} \geq 0 \text{ for all } i = 1, 2, 3, 4, 5 \text{ and } j = 1, 2, 3, 4, 5$$
$$and\ binary\ \{0,1\}$$

We will state there are many methods to use such as the Hungarian method or integer programming. Since our Table 4.1 is all integers, we may use a similar technique as we did in the transportation problems in Chapter 3 (see Albright and Fox, 2020).

Our solution is that the minimum miles is 28 (in hundreds) when I4 is assigned to A, I3 assigned to B, I1 is assigned to C, I2 is assigned to D, and I5 is assigned to E.

4.1.1 Assignment Problem

LP Formulation:

Min $10X_{11} + 4X_{12} + 6X_{13} + 10X_{14} + 12X_{15} + 11X_{21} + 7X_{22} + 7X_{23} + 9X_{24} + 14X_{25}$
$+ 13X_{31} + 8X_{32} + 12X_{33} + 14X_{34} + 15X_{35} + 14X_{41} + 16X_{42} + 13X_{43} + 17X_{44} +$
$17X_{45} + 19X_{51} + 11X_{52} + 17X_{53} + 20X_{54} + 19X_{55}$

subject to

$$X_{11} + X_{12} + X_{13} + X_{14} + X_{15} = 1$$
$$X_{21} + X_{22} + X_{23} + X_{24} + X_{25} = 1$$
$$X_{31} + X_{32} + X_{33} + X_{34} + X_{35} = 1$$
$$X_{41} + X_{42} + X_{43} + X_{44} + X_{45} = 1$$
$$X_{51} + X_{52} + X_{53} + X_{54} + X_{55} = 1$$
$$X_{11} + X_{21} + X_{31} + X_{41} + X_{51} = 1$$
$$X_{12} + X_{22} + X_{32} + X_{42} + X_{52} = 1$$
$$X_{13} + X_{23} + X_{33} + X_{43} + X_{53} = 1$$
$$X_{14} + X_{24} + X_{34} + X_{44} + X_{54} = 1$$
$$X_{15} + X_{25} + X_{35} + X_{45} + X_{45} = 1$$
$$X_{ij} \geq 0$$

After running the assignment problem on any LP Solver or Excel, the results are:

Person 1 should do job 3. Person 2 should do job 4. Person 3 should do job 5.

Person 4 should do job 1. Person 5 should do job 2.

The total cost is $55.

4.2 EXERCISES

4.1. Let us consider the case of a Fix-It-Shop, which has just received three new rush projects to repair: (1) A radio, (2) a toaster oven, and (3) a broken coffee table. Three repair people, each with different talents and abilities, are available to do the jobs. The owner of the shop estimates what it will cost in wages to assign each of the workers to each of the three projects. The costs, which are shown in Table 4.2, differ because the owner believes that each worker will differ in speed and skill on these quite varied jobs. Table 4.2 summarizes all six assignment options. The table also shows that the least-cost solution would be to assign Cooper to project 1, Brown to project 2, and Adams to project 3, at a total cost of $25. The owner's objective is to assign the three projects to the workers in a way that will result in the lowest cost to the shop. Note that the assignment of people to projects must be on a one-to-one basis; each project will be assigned exclusively to one worker only.

TABLE 4.2 Fix-It-Shop Repair Team Wages

| | Project | | |
Person	1	2	3
Adams	$11	$14	$6
Brown	8	10	11
Cooper	9	12	7

TABLE 4.3 Accountant Data for Exercise 4.2

	1	2	3
Phil	15	9	12
Joy	7	5	10
Ralph	13	4	6

TABLE 4.4   Shawn Taxi Data for Exercise 4.3

	P	Q	R	S
1	10	8	4	6
2	6	4	12	8
3	14	8	10	2
4	4	14	10	8

4.2. Three accountants, Phil, Joy, and Ralph, are to be assigned three tax projects 1, 2, and 3. The assignment costs are given in Table 4.3.

4.3. Shawn's Taxi has four taxis, 1, 2, 3, and 4, and there are four customers that called in for services P, Q, R, and S. The distances between the taxis and the customers' locations are (given in miles) in Table 4.4. The Taxi company wishes to assign the taxis to customers so that the distance traveled is minimized.

REFERENCES AND ADDITIONAL READINGS

Albright, B. and W.P. Fox. (2020). *Mathematical Modeling with Excel*. 2nd Ed. Taylor and Francis, CRC Press.

CHAPTER 5

Mathematical Programming Methods

THIS CHAPTER WILL INTRODUCE you to several additional mathematical programming methods that include Data Envelopment Analysis (DEA), flow problems, and integer programming problems. Each of the methods presented in this chapter will support the OR practitioner in addressing a range of business-related problems.

5.1 DATA ENVELOPMENT ANALYSIS (DEA)

Data Envelopment Analysis is a nonparametric method in operations research and economics for the estimation of production activities. Practitioners have applied DEA in a large range of areas including banking, economic sustainability, police/fire department operations, and logistical problems. DEA is also gaining popularity in assessing the performance of natural language processing models and within machine learning.

Consider a manufacturing process where the manager thinks that at least one or more of the departments needs improving to do a better overall job. Rank ordering these departments is a good method to see how each department is doing. We will use data envelopment analysis to help with this problem.

Data envelopment analysis (DEA) is a data input–output-driven approach for evaluating the performance or efficiency of entities called decision-making units (DMUs). This method relates the measures of inputs to the measures of outputs. The definition of a DMU is generic and

DOI: 10.1201/9781003464969-5

very flexible so that any entity to be ranked might be a DMU. DEA has been used to evaluate the *performances* or *efficiencies* of hospitals, schools, departments, US Air Force wings, US armed forces recruiting agencies, universities, cities, courts, businesses, banking facilities, countries, regions, SOF airbases, key nodes in networks, and the list goes on. DEA has been used to gain insights into activities that were not obtained by other quantitative or qualitative methods.

DEA is described as a mathematical programming model applied to observational data. It provides a new way of obtaining empirical estimates of relationships among the DMUs.

The model, in its simplest terms, may be formulated and solved as a linear programming problem (Winston, 1995; Trick, 1996). Although several formulations for DEA exist, we seek the most straightforward formulation to maximize the efficiency of a DMU as constrained by their inputs and outputs as shown in Equation 5.1. As an option, we might normalize the metric inputs and outputs for the alternatives if the values are poorly scaled within the data. We will call this data matrix $\mathbf{X}$, with entries x_{ij}. We define an efficiency unit as E_i for $I = 1, 2, \ldots, nodes$. We let w_i be the weights or coefficients for the linear combinations. Further, we restrict any efficiency from being larger than one. Thus, the largest efficient DMU will be 1 corresponding to 100%. This gives the following linear programming formulation for single outputs but multiple inputs:

$$Max\, E_i$$

subject to

$$\sum\nolimits_{i=1}^{n} w_i x_{ij} - E_i = 0, j = 1, 2, \ldots \qquad \text{(Eq. 5.1)}$$

$$E_i \leq 1 \text{ for all } i$$

For multiple inputs and outputs, we recommend the formulations provided by Winston (1995) and Trick (2014) using Equation 5.1.

For any DMU_0, let X_i be the inputs and Y_i be the outputs. Let X_0 and Y_0 be the DMU modeled.

$$Min \ \theta$$

Subject to:

$$\Sigma \lambda i X i < \theta X0 \qquad \text{(Eq. 5.2)}$$
$$\lambda i Y i < Y0$$
$$\lambda \iota > 0 \ Non\text{-}negativity$$

DEA can be a very useful tool when used wisely according to Trick (2012). A few of the strengths that make DEA extremely useful are (Trick, 2012):

(1) DEA can handle multiple input and multiple output models.

(2) DEA doesn't require an assumption of a functional form, relating inputs to outputs.

(3) DMUs are directly compared against a peer or combination of peers.

(4) Inputs and outputs can have very different units. For example, X_1 could be in units of lives saved, and X_2 could be in units of dollars without requiring any a priori trade-off between the two.

The same characteristics that make DEA a powerful tool can also create limitations to the process and analysis. An analyst should keep these limitations in mind when choosing whether or not to use DEA. A few additional limitations include:

(1) Since DEA is an extreme point technique, noise in the data such as measurement error can cause significant problems.

(2) DEA is good at estimating "relative" efficiency of a DMU, but it converges very slowly to "absolute" efficiency. In other words, it can tell you how well you are doing compared to your peers but not compared to a "theoretical maximum".

(3) Since DEA is a nonparametric technique, statistical hypothesis tests are difficult and are the focus of ongoing research.

(4) Since a standard formulation of DEA with multiple inputs and outputs creates a separate linear program for each DMU, large problems can be computationally intensive.

(5) Linear programming does not ensure all weights are considered. We find that the values for weights are only for those that optimally determine an efficiency rating. If having all criteria weighted (inputs, outputs) is essential to the decision-maker, do not use DEA.

5.1.1 Sensitivity Analysis

Sensitivity analysis is always an important element in analysis. According to Neralic (1998), an increase in any output cannot make a solution worse rating nor can a decrease in inputs alone worsen an already achieved efficiency rating. As a result, in our examples, we only decrease outputs and increase inputs as just described (Neralic, 1998). We will illustrate some sensitivity analysis, as applicable, in our illustrious examples later.

5.2 MANUFACTURING PROBLEM WITH DEA

Consider the following manufacturing process, (Winston, 1995 and Trick, 2014) where we have three DMUs each of which has 2 inputs and 3 outputs as shown in the data (Table 5.1).

Since no units are given and the scales are similar, we decide not to normalize the data. We define the following decision variables:

Decision variables:

t_i = Value of a single unit of output of DMU_i, for $i = 1, 2, 3$
w_i = Cost or weights for one unit of inputs of DMU_i, for $i = 1, 2$

$efficiency_i = DMU_i$ = (Total value of i's outputs)/(Total cost of i's inputs), for $i = 1, 2, 3$

The following modeling assumptions are made:

1. No DMU will have an efficiency of more than 100%.

2. If any efficiency is less than 1, then it is inefficient.

TABLE 5.1   Manufacturing Problem Data

DMU	Input#1	Input#2	Output#1	Output#2	Output#3
1	5	14	9	4	16
2	8	15	5	7	10
3	7	12	4	9	13

3. We will scale the costs so that the costs of the inputs equals 1 for each linear program. For example, we will use $5w_1 + 14w_2 = 1$ in our program for DMU_1.

4. All values and weights must be strictly positive, so we use a constant such as 0.0001 in lieu of 0.

To calculate the efficiency of DMU_1, we define the linear program using Equation 5.2 as

objective function:

$$\text{Maximize } DMU_1 = 9t_1 + 4t_2 + 16t_3$$

Subject to:

$$- 9t_1 - 4t_2 - 16t_3 + 5w_1 + 14w_2 \geq 0$$

$$- 5t_1 - 7t_2 - 10t_3 + 8w_1 + 15w_2 \geq 0$$

$$- 4t_1 - 9t_2 - 13t_3 + 7w_1 + 12w_2 \geq 0$$

$$5w_1 + 14w_2 = 1$$

$$t_i \geq 0.0001, i = 1, 2, 3$$

$$w_i \geq 0.0001, i = 1, 2$$

Non-negativity

To calculate the efficiency of DMU_2, we define the linear program using Equation 5.2 as

objective function:

$$\text{Maximize } DMU_2 = 5t_1 + 7t_2 + 10t_3$$

Subject to:

$$- 9t_1 - 4t_2 - 16t_3 + 5w_1 + 14w_2 \geq 0$$

$$- 5t_1 - 7t_2 - 10t_3 + 8w_1 + 15w_2 \geq 0$$

$$- 4t_1 - 9t_2 - 13t_3 + 7w_1 + 12w_2 \geq 0$$

$$8w_1 + 15w_2 = 1$$

$$t_i \geq 0.0001, i = 1, 2, 3$$

$$w_i \geq 0.0001, i = 1, 2$$

Non-negativity

To calculate the efficiency of DMU_3, we define the linear program as

objective function:

$$\text{Maximize } DMU_3 = 4t_1 + 9t_2 + 13t_3$$

Subject to:

$$-9t_1 - 4t_2 - 16t_3 + 5w_1 + 14w_2 \geq 0$$

$$-5t_1 - 7t_2 - 10t_3 + 8w_1 + 15w_2 \geq 0$$

$$-4t_1 - 9t_2 - 13t_3 + 7w_1 + 12w_2 \geq 0$$

$$7w_1 + 12w_2 = 1$$

$$t_i \geq 0.0001, i = 1, 2, 3$$

$$w_i \geq 0.0001, i = 1, 2$$

Non-negativity

The linear programming solutions show the efficiencies as $DMU_1 = DMU_3 = 1$, $DMU_2 = 0.77303$.

Interpretation: DMU_2 is operating at 77.303% of the efficiency of DMU_1 and DMU_3. Management could concentrate on some improvements or best practices from DMU_1 or DMU_3 for DMU_2. An examination of the dual prices for the linear program of DMU_2 yields $\lambda 1 = 0.261538$, $\lambda 2 = 0$, and $\lambda 3 = 0.661538$. The average output vector for DMU_2 can be written as:

$$0.261538 \begin{bmatrix} 9 \\ 4 \\ 16 \end{bmatrix} + 0.661538 \begin{bmatrix} 4 \\ 9 \\ 13 \end{bmatrix} = \begin{bmatrix} 5 \\ 7 \\ 12.785 \end{bmatrix}$$

and the average input vector can be written as:

$$0.261538 \begin{bmatrix} 5 \\ 14 \end{bmatrix} + 0.661538 \begin{bmatrix} 7 \\ 12 \end{bmatrix} = \begin{bmatrix} 5.938 \\ 11.6 \end{bmatrix}.$$

In our data, output #3 is 10 units. Thus, we may clearly see the inefficiency is in output #3 where 12.785 units are required. We find that they are short 2.785 units (12.785−10 = 2.785). This helps focus on treating the inefficiency found for output #3.

Sensitivity analysis: Sensitivity analysis in a linear program is sometimes referred to as "what if" analysis. Let's assume that without management engaging some additional training for DMU_2 that DMU_2 output #3 dips from 10 to 9 units of output, while the input 2 hours increases from 15 to 16 hours. We find that these changes in the *technology coefficients* are easily handled in resolving the LPs. Since DMU_2 is affected, we might only modify and solve the LP concerning DMU_2. We find with these changes that DMU's efficiency is now only 74% as effective as DMU_1 and DMU_3.

Conclusion: DEA is a good method for ranking entities. We have successfully used it to rank schools and departments within colleges, banks, and recruiting offices for the armed forces.

We will follow DEA with a discussion of flow-related problems, including shortest path, max flow, and identifying critical flow problems.

5.3 SHORTEST PATH PROBLEMS

The shortest path problem, as the name implies, involves finding the shortest path between two vertices (or nodes) in a graph. Applications of the shortest path problem include those in developing highway networks, logistics and communications networks, electronic component design, and power grid analysis.

Given a number of cities with highways connecting them, for example, we need to find the shortest path from New York to Chicago.

The traffic and length of the highways are path weights.

First, we provide three algorithms that may be used to determine the shortest path between items: Kruskal, Prims, and Dykstra.

Why shortest path problems? Shortest path algorithms are a family of algorithms designed to solve the shortest path problems. The shortest path problem is something most people have some intuitive familiarity with: Given two points, A and B, what is the shortest path between them? In computer science, however, the shortest path problem can take different forms, and so different algorithms are needed to be able to solve them all.

Shortest path algorithms have many applications. Mapping software like Google or Apple maps make use of shortest path algorithms. They are also important for road network, operations, and logistics research. Shortest path algorithms are also very important for computer networks, like the internet.

For simplicity and generality, shortest path algorithms typically operate on some input graph, G. This graph is made up of a set of vertices, V, and edges, E, that connect them. If the edges have weights, the graph is called a weighted graph. Sometimes, these edges are bidirectional, and the graph is called undirected. Sometimes, there can be even be cycles in the graph. Each of these subtle differences are what makes one algorithm work better than another for certain graph type.

Let us start with an example like the network in Figure 5.1.

5.3.1 Network Analysis

Given the following network (in Figure 5.1), determine the shortest path.

We will provide the algorithms for each method and the results for this example.

5.3.2 Kruskal's Algorithm

INPUT: a weighted graph with n vertices

Body: 1. Initialize a graph T with no edges.

  2. Let E be the set of all edges of the graph.

  3. While ($m < n$-1)

    3a. Find the edge e in E that has minimum weight.

    3b. Delete e from set E

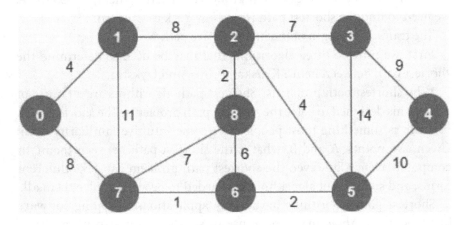

FIGURE 5.1 Network with 9 Nodes and 14 Edges.

3c. If adding *e* to T does not produce a closed circuit (loop) then add *e* to T and update both T and E.

OUTPUT: optimal graph and minimum weight.

5.3.3 Kruskal's Method for Network Analysis Problem

Nodes {0, 1, 2, 3, 4, 5,6,7,8}

Edges {0–1, 0–7, 1–8, 1–7, 7–8, 7–6, 1–2, 2–8, 2–3, 2–5, 6–5, 3–5, 3–4, 5–4}

In this graph, the edges go to both directions.

We put the edges in order from smallest to largest distance (weight).

Edge	Distance
7–6	1
6–5	2
2–8	3
0–1	4
2–5	5
6–8	6
7–8	7
2–3	8
0–7	9
1–2	10
3–4	11
5–4	12
1–7	13
3–5	14

We start with the smallest edge 7–6 with value 1.

Next, we have a tie between 2 and 8 or 6 and 5 with value 2. We arbitrarily choose 6–5 with value 2. Next, we choose 2–8 also with value 2.

Next, we have ties between 0 and 1 at 4 and 2 and 5 at 4. We arbitrarily choose 2–5 at 4 first and then we see we can also add 0–1 at 4 because none of these cause a closed loop.

Next, we have 6–8 and see that including it causes a closed loop, so we discard it and move on. We see 7–8 also causes a closed loop, but 2–3 does not. We add 2–3 to our solution with value 7. Next, we add 0–7 with value 8 and see 7–8 cannot be added. Next, we add edges 3–4 with value 9.

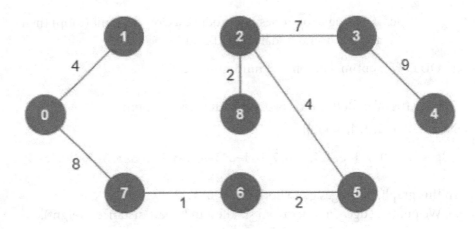

MST using Kruskal's algorithm

FIGURE 5.2   Minimal Spanning Tree by Kruskal's Algorithm with Total Value of 37.

The solution, with any closed loops, is shown in Figure 5.2.

5.3.4 Prim's Algorithm

INPUT: a weighted graph with n vertices

BODY: 1. Pick a vertex v and put into set C

　　2. Let set C' be the set of all vertices except v.

　　3. For i=1 to n-1

　　　　3a. Find an edge from C' to connect to C that has a

　　　　minimum weight and does not form a closed loop.

　　　　3b. Add that edge to C and update both C and C'.

OUTPUT: Optimal graph and minimum weight.

We arbitrarily choose edge 0–1 at value 4 to start,

We can now choose either 1–2 with value 8, 0–7 at value 8, or 1–7 at value 11. The smallest is a tie between 0 and 7 and 1 and 2 each with value 8. We choose arbitrarily 1–2 at value 8. Currently, we have 0–1–2 with total value 12.

Choices are now:

$$0\text{-}7 \ \{8\}$$

$$1\text{-}7 \ \{11\}$$

$$2\text{-}8 \ \{2\}$$

The smallest is 2–8 {2}, and it does not cause a closed loop. We currently have {0–1, 1–2, 2–8} with a total value of 14.

Choices are:

$$0\text{-}7 \ \{8\}$$

$$1\text{-}7 \ \{11\}$$

$$2\text{-}5 \ \{4\}$$

$$2\text{-}3 \ \{7\}$$

$$7\text{-}8 \ \{7\}$$

$$8\text{-}6 \ \{6\}$$

The smallest is 2–5 {4} which does not close a loop. We have edges {0–1, 1–2, 2–8, 2–5} with value 18.

Choices are:

$$0\text{-}7 \ \{8\}$$

$$1\text{-}7 \ \{11\}$$

$$2\text{-}3 \ \{7\}$$

$$7\text{-}8 \ \{7\}$$

$$8\text{-}6 \ \{6\}$$

$$5\text{-}4 \ \{10\}$$

$$5\text{-}6 \ \{2\}$$

$$5\text{-}3 \ \{14\}$$

The smallest is 5–6 {2} which does not close a loop. We add 5–6 to our edges {0–1, 1–2, 2–8, 2–5, 5–6} with total value 20.

Choices are:

$$0\text{--}7\ \{8\}$$

$$1\text{--}7\ \{11\}$$

$$2\text{--}3\ \{7\}$$

$$7\text{--}8\ \{7\}$$

$$8\text{--}6\ \{6\}$$

$$5\text{--}4\ \{10\}$$

$$5\text{--}3\ \{14\}$$

$$6\text{--}7\ \{1\}$$

The smallest is 6–7 {1} which does not close a loop. We add {0–1, 1–2, 2–8, 2–5, 5–6, 6–7} to our edges with total value 21.

Choices are:

$$0\text{--}7\ \{8\}$$

$$1\text{--}7\ \{11\}$$

$$2\text{--}3\ \{7\}$$

$$7\text{--}8\ \{7\}\ \text{Closed loop}$$

$$8\text{--}6\ \{6\}$$

$$5\text{--}4\ \{10\}$$

$$5\text{--}3\ \{14\}$$

$$7\text{--}1\ \{11\}\ \text{Closed loop}$$

The smallest is 2–3 {7} which does not close a loop. We add 2–3 to our edges {0–1, 1–2, 2–8, 2–5, 5–6, 6–7, 2–3} with total value 28.

Choices are:

$$0\text{--}7\ \{8\}\ \text{Closed loop}$$

$$1\text{--}7\ \{11\}$$

$$3\text{--}4\ \{9\}$$

3–5 {14}

7–8 {7} Closed loop

8–6 {6} Closed loop

5–4 {10}

5–3 {14}

7–1 {11} Closed loop

The smallest is 3–4 {9} which does not close a loop. We add 3–4 to our edges {0–1, 1–2, 2–8, 2–5, 5–6, 6–7, 2–3, 3–4} with total value 37.
We have completed our MST.

5.3.5 Dijkstra's Algorithm

1) Create a set *sptSet* (shortest path tree set) that keeps track of vertices included in shortest path tree, i.e., whose minimum distance from source is calculated and finalized. Initially, this set is empty.

2) Assign a distance value to all vertices in the input graph. Initialize all distance values as INFINITE. Assign distance value as 0 for the source vertex so that it is picked first.

3) While *sptSet* doesn't include all vertices,

   a) Pick a vertex u which is not there in *sptSet* and has minimum distance value.

   b) Include u to *sptSet*.

   c) Update distance values of all adjacent vertices of u. To update the distance values, iterate through all adjacent vertices. For every adjacent vertex v, if the sum of distance value of u (from source) and weight of edge u-v is less than the distance value of v, then update the distance value of v.

Nodes {0, 1, 2, 3, 4, 5,6,7,8}
   Start at node 0 (distance is 0).
   Go to node 1 with distance = 4 or node 7 distance is = 8. Node 1 is chosen 0–1, total distance = 8.
   We continue in this fashion until we complete the tree with value 37.

5.4 MAXIMUM FLOW PROBLEM

In maximum flow problems, we want to obtain the maximum flows (numbers) through a network. In the maximum flow problem, we have a directed graph with a source node s and a final node t, and each edge has a capacity that represents the maximum amount of flow that can be sent through it. The goal is to find the maximum amount of flow that can be sent from s to t, while respecting the capacity constraints on the edges.

5.4.1 Example 5.1: Maximum Flow through a Given Network

We want to maximize the flow from the source at Node 1 to the sink ending at Node 7, passing through the other Arcs whose capacities are indicated.

Applications: – Traffic movement – Hydraulic systems – Electrical circuits – Layouts

Source/Sink	1	2	3	4	5	6	7
1	—	12	12				
2			1	8		6	
3		1			12	1	
4						2	7
5						2	8
6				4	2	3	2
7							

After solving this LP problem using say LINDO, the results are as follows:

Max F

Subject to:

$$x_{12}+x_{13}-F<0$$

$$x_{12}+x_{13}-F>0$$

$$x_{12}+x_{32}-x_{23}-x_{26}-x_{24}<0$$

$$x_{12}+x_{32}-x_{23}-x_{26}-x_{24}>0$$

$$x_{13}+x_{23}+x_{63}-x_{32}-x_{36}-x_{35}<0$$

$$x_{13}+x_{23}+x_{63}-x_{32}-x_{36}-x_{35}>0$$

$$x_{24}+x_{64}-x_{47}-x_{46}<0$$

$$x_{35}+x_{65}-x_{36}-x_{57}<0$$

$$x_{26}+x_{46}+x_{36}+x_{56}-x_{65}-x_{63}-x_{64}-x_{67}<0$$

$$x_{47}+x_{67}+x_{57}-F<0$$

$$x_{24}+x_{64}-x_{47}-x_{46}>0$$

$$x_{35}+x_{65}-x_{36}-x_{57}>0$$

$$x_{26}+x_{46}+x_{36}+x_{56}-x_{65}-x_{63}-x_{64}-x_{67}>0$$

$$x_{47}+x_{67}+x_{57}-F>0$$

$$x_{12}<12$$

$$x_{13}<12$$

$$x_{23}<1$$

$$x_{32}<1$$

$$x_{26}<6$$

$$x_{36}<4$$

$$x_{63}<4$$

$$x_{24}<8$$

$$x_{64}<3$$

$$x_{46}<3$$

$$x_{35}<12$$

$$x_{65}<2$$

$$x_{57}<8$$

$$x_{47}<7$$

$$x_{67}<2$$

end

LP OPTIMUM FOUND AT STEP 2

```
OBJECTIVE FUNCTION VALUE

1)  17.00000
```

VARIABLE	VALUE	REDUCED COST
F	17.000000	0.000000
X_{12}	12.000000	0.000000
X_{13}	5.000000	0.000000
X_{32}	0.000000	0.000000
X_{23}	0.000000	0.000000
X_{26}	5.000000	0.000000
X_{24}	7.000000	0.000000
X_{63}	1.000000	0..000000
X_{36}	0.000000	0.000000
X_{35}	6.000000	0.000000
X_{64}	0.000000	0.000000
X_{47}	7.000000	0.000000
X_{46}	0.000000	0.000000
X_{65}	2.000000	0.000000
X_{57}	8.000000	0.000000
X_{56}	0.000000	0.000000
X_{67}	2.000000	0.000000

5.5 CRITICAL PATH IN PROJECT PLAN NETWORK

The successful management of large projects, whether they are construction, transportation, or financial, relies on careful scheduling and coordinating of various tasks. Critical Path Method (CPM) attempts to analyze project scheduling. This allows for better control and evaluation of the project. For example, we want to know how long will the project take? When will we be able to start a particular task? If this task is not completed on time, will the entire project be delayed? Which tasks should we speed up (crash) in order to finish the project earlier?

Given a network of activities, the first problem of interest is to determine the length of time required to complete the project and the set of critical activities that control the project completion time. Suppose that in a given project activity network, there are m nodes, n arcs (i.e., activities), and an estimated duration time, C_{ij}, associated with each arc (i to j) in the network. The beginning node of an arc corresponds to the start of the associated activity and the end node to the completion of an activity. To find the Critical Path (CP), define the binary variables X_{ij}, where $X_{ij} = 1$

if the activity $i\,j$ is on the CP and $X_{ij} = 0$ otherwise. The length of the path is the sum of the duration of the activities on the path. The length of the longest path is the shortest time needed to complete the project. Formally, the CP problem is to find the longest path from node 1 to node m.

Each arc has two roles: It represents an activity, and it defines the precedence relationships among the activities. Sometimes, it is necessary to add arcs that only represent precedence relationships. These *dummy arcs* are represented by dashed arrows. In our example, the arc from 2 to 3 represents a dummy activity.

The first constraint says that the project must start. For each intermediate node, if we ever reach it we have to leave that node. Finally, the last constraint enforces the completion of the project.

5.5.1 Example 5.2: CPM

Critical Path Method

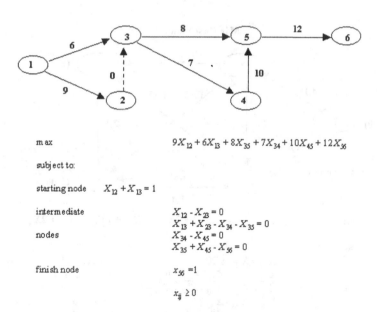

max
$$9X_{12} + 6X_{13} + 8X_{35} + 7X_{34} + 10X_{45} + 12X_{56}$$

subject to:

starting node $\quad X_{12} + X_{13} = 1$

intermediate
$$X_{12} - X_{23} = 0$$
$$X_{13} + X_{23} - X_{34} - X_{35} = 0$$
nodes
$$X_{34} - X_{45} = 0$$
$$X_{35} + X_{45} - X_{56} = 0$$

finish node $\quad X_{56} = 1$

$$x_{ij} \geq 0$$

Running the LP formulation on any LP Solver, the critical path is:

From node 1 to 2

From node 2 to 3

From node 3 to 4

From node 4 to 5

From node 5 to 6

The duration of the project is found to be 38 time units.

5.6 MINIMUM COST FLOW PROBLEM

All the above network problems are special cases of the minimum cost flow problem. Like the maximum flow problem, it considers flows in networks with capacities. Like the shortest path problem, it considers a cost for flow through an arc. Like the transportation problem, it allows multiple sources and destinations. Therefore, all of these problems can be seen as special cases of the minimum cost flow problem.

The problem is to minimize the total cost subject to availability and demand at some nodes and upper bound on flow through each arc.

5.6.1 Example 5.3: Minimum Cost Flow through a Network

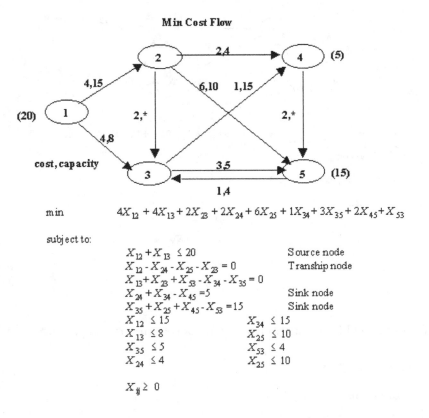

Min Cost Flow

$$\min \quad 4X_{12} + 4X_{13} + 2X_{23} + 2X_{24} + 6X_{25} + 1X_{34} + 3X_{35} + 2X_{45} + X_{53}$$

subject to:

$$X_{12} + X_{13} \leq 20 \qquad \text{Source node}$$
$$X_{12} - X_{24} - X_{25} - X_{23} = 0 \qquad \text{Tranship node}$$
$$X_{13} + X_{23} + X_{53} - X_{34} - X_{35} = 0$$
$$X_{24} + X_{34} - X_{45} = 5 \qquad \text{Sink node}$$
$$X_{35} + X_{25} + X_{45} - X_{53} = 15 \qquad \text{Sink node}$$

$$
\begin{array}{ll}
X_{12} \leq 15 & X_{34} \leq 15 \\
X_{13} \leq 8 & X_{25} \leq 10 \\
X_{35} \leq 5 & X_{53} \leq 4 \\
X_{24} \leq 4 & X_{25} \leq 10
\end{array}
$$

$$X_{ij} \geq 0$$

The optimal solution is: $X_{12} = 12$, $X_{13} = 8$, $X_{23} = 8$, $X_{24} = 4$, $X_{34} = 11$, $X_{35} = 5$, $X_{45} = 10$, all other $X_{ij} = 0$. The optimal cost is $150.

5.7 GENERAL INTEGER LINEAR PROGRAMS

Standard LP assumes that decision variables are continuous. However, in many applications, fractional values may be of little use (e.g., 2.5 employees). On the other hand, as you know by now, since integer linear programs are more difficult to solve, you might ask why bother. Why not simply use a standard linear program and round the answers to the nearest integers? Unfortunately, there are two problems with this:

- The rounded solution may be infeasible.

- Rounding may not give an optimal solution.

Therefore, rounding the results from linear programs can give reasonable answers, but to guarantee optimal solutions, we have to use integer linear programming. By default, LP software assumes that all variables are continuous. In using LINDO software, you will want to make use of the general integer statement – GIN. GIN followed by a variable name restricts the value of the variable to the nonnegative integers (0,1,2, . . .). The following small example illustrates the use of the GIN statement.

Doing Integer LP by *Excel Solver*: Using the Constraint menu, for the LP Problem, select Normal Constraint and then the icon for "< = " to obtain the Add Constraint window. Then we can designate any variable (e.g., enter the variable, say $B$5 in the Cell Reference) as Integer in the Add Constraint window.

5.7.1 Example 5.4: Manufacturing Equipment

We manufacture two sporting pieces of equipment: Baseballs and softballs. Our formulation with x_1 being baseballs and x_2 being softballs is using LINDO:

```
Max 11X₁ + 10X₂
S.T.  2X₁ + X₂   ≤ 12
          X₁ - 3X₂  ≥ 1
END
GIN X₁
GIN X₂
```

The output after seven iterations is:

```
OBJECTIVE FUNCTION VALUE

1)      66.00000

VARIABLE         VALUE              REDUCED COST
   X₁           6.000000            -11.000000
   X₂           0.000000            -10.000000


ROW    SLACK OR SURPLUS
 2)         0.000000
 3)         5.000000
```

Had we not specified X_1 and X_2 to be general integers in this model, LINDO would not have found the optimal solution of $X_1 = 6$ and $X_2 = 0$. Instead, LINDO would have treated X_1 and X_2 as continuous and returned the solution of $X_1 = 5.29$ and $X_2 = 1.43$.

Note also that simply rounding the continuous solution to the nearest integer values does not yield the optimal solution in this example. In general, rounded continuous solutions may be nonoptimal and, at worst, infeasible. Based on this, one can imagine that it can be very time consuming to obtain the optimal solution to a model with many integer variables. In general, this is true, and you are best off utilizing the GIN feature only when absolutely necessary.

As a final note, the GIN command also accepts an integer value argument in place of a variable name. The number corresponds to the number of variables you want to be general integers. These variables must appear first in the formulation. Thus, in this simple example, we could have replaced our two GIN statements with the single statement: GIN 2.

Most commercial solvers use the "branch and bound" or "branch and cut" method in solving IP. The first step is usually is to relax the IP to an LP problem. However, some software, such as CPLEX will then automatically execute heuristic methods to round the LP solution to get a good (and feasible) solution to the IP. The difference between you doing this yourself and letting an IP solver do it is that the IP solver is likely to consider the payback constraint sacrosanct and reject any rounded solution that violates it.

5.7.2 Example 5.5: Integer LP Programs by Excel

Suppose we have a manufacturing problem to produce tables and chairs. In our problem, every table, x_2, needs four chairs, x_1; then the LP formulation is:

Max $5X_1 + 3X_2$

S.T.
$2X_1 + X_2 \leq 40$
$X_1 + 2X_2 \leq 50$
$4X_1 - X_2 = 0$
$X_1 \geq 0 \quad X_2 \geq 0$

The setup in Excel:

	Guess		
x_1	0		
x_2	0		
OBJ FUNC	0		
Constraints			Available
	0		40
	0		50
	0		0

Initial LP solution:

	Guess		
x_1	5.56		
x_2	22.22		
OBJ FUNC	94.44		
Constraints			Available
	33.33		40
	50		50
	0		0

We need integer solutions, so we modify the LP in Excel.

	Guess		
X_1	5		
X_2	20		
OBJ FUNC	85		
Constraints			Available
	30		40
	45		50
	0		0

We solve to obtain the integer solutions:

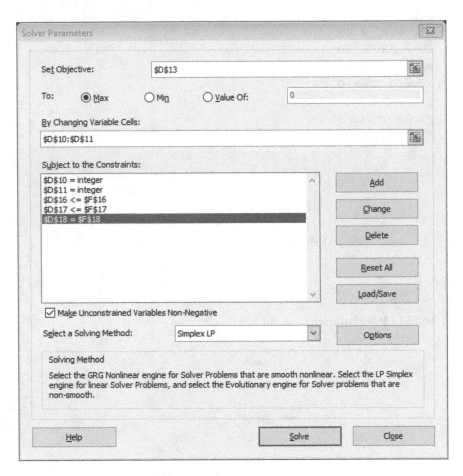

The optimal solution is ($X_1 = 5$, $X_2 = 20$) with optimal value of 85.

5.8 MIXED INTEGER PROGRAMMING APPLICATION: "EITHER-OR" CONSTRAINTS

Suppose a bakery sells eight varieties of doughnuts. The preparation of varieties 1, 2, and 3 involves a rather complicated process, and so the bakery has decided that it would rather not bake these varieties unless it can bake and sell at least ten dozen doughnuts of varieties 1, 2, and 3 combined. Suppose also that the capacity of the bakery prohibits the total number of doughnuts baked from exceeding 30 dozens, and that the per unit profit for a variety j doughnut is P_j dollars. If we let X_j, $j = 1, 2, \ldots, 6$ denote the number of dozens of doughnut of variety j to be baked, then the maximum profit can be found by solving the following problem (assuming the bakery can sell everything it bakes):

$$\text{Maximize } Z = \Sigma\, P_j\, X_j$$

$$\text{Subject to the constraints: } \Sigma\, X_j \leq 30$$

$$X_1 + X_2 + X_3 = 0, \text{ or, } X_1 + X_2 + X_3 \geq 10$$

Where all variables are nonnegative. The above "either-or" constraint can be handled as such: Let y be a 0–1 variable. Then an equivalent problem is:

$$\text{Maximize } Z = \Sigma\, Pj\, Xj$$

$$\text{Subject to: } \Sigma\, Xj \leq 30$$

$$X_1 + X_2 + X_3 - 30y \leq 0,$$

$$X_1 + X_2 + X_3 - 10y \geq 10$$

$$Xj \geq 0,\, y = 0, \text{ or } 1.$$

5.8.1 Conditional Relations among Constraints

Suppose that, in order for a certain variable X to be positive, it is necessary that another variable Y exceed a certain threshold value. This conditional statement can be reformulated as:

$$(x = 0) \text{ OR } (y \geq a).$$

Since a conditional statement can be expressed as an either-or statement by negating the hypotheses, the either-or statement can be expressed as:

$$X <= d{*}M,\, y >= a{*}d,\, 0 <= d <= 1, \text{ where } d \text{ is an integer.}$$

Here M is a very large positive number.

Off-On Constraints: Suppose we wish a variable to take on the value a or else to be 0. It is easy to accomplish this by means of the following conditions: $X = da$; $0 \leq d \leq 1$ where d is an integer.

Clearly, that if d satisfies the last two conditions, the only possible values it can take on are 0 and 1. So if $d = 0$, then $x = 0$, and if $d = 1$, then $X = a$, as required.

Off-on intervals: Suppose we want x either to be 0 or else to be in a fixed interval between a and b. The following inequalities accomplish these requirements: $da \leq X \leq db; 0 < d < 1$ and d integer.

Clear that if $d = 0$, then $x = 0$, and if $d = 1$, then x satisfies $a \leq X \leq b$. These are the only possible values for d.

5.8.2 A Case of Discrete Finite Valued Variables

Suppose a variable X takes only one of a finite number of values $a_1, a_2, \ldots$ and a_n. Set:

$$X = a_1 d_1 + a_2 d_2 + \ldots + a_n d_n, \Sigma d_i = 1, 0 < d_i < 1 \text{ and } d_i \text{ is an integer.}$$

Notice that the conditions on the d_i's require that exactly one of them should be 1 and the rest 0. And if $d_i = 1$, then $X = a_i$.

5.8.3 0–1 Integer Linear Programs

Suppose there are n items to be considered for inclusion in a knapsack. Each item has certain per unit value to the traveler who is packing the knapsack. Each item has a per unit weight that contributes to the overall weight of the knapsack. There is a limitation on the total weight that can be carried. The objective is to maximize the total value of what is packed into the knapsack subject to the total weight limitation. We can use Binary LP to solve this problem.

Using the INT command in LINDO restricts a variable to being either 0 or 1. These variables are often referred to as binary variables. In many applications, binary variables can be very useful in modeling all-or-nothing situations. Examples might include such things as taking on a fixed cost, building a new plant, or buying a minimum level of some resource to receive a quantity discount.

5.9 ILLUSTRIOUS EXAMPLES

5.9.1 Example 5.7: Knapsack Problem

$$\text{Maximize } 11X_1 + 9X_2 + 8X_3 + 15X_4$$

$$\text{Subject to: } 4X_1 + 3X_2 + 2X_3 + 5X_4 \leq 8, \text{ and any } Xi \text{ is either 0 or 1.}$$

Since this is a very small size problem, there are four variables and each can have either of two values, there are $2^4 = 16$ possibilities. Trying all 16 possibilities in order to identify an optimal solution (if it exists) is tedious. Therefore, one must use any one of ILP software packages to solve even this or any larger-scale problem.

Using LINDO, the problem statement is:

```
Max  11X₁ +  9X₂ +  8X₃ + 15X₄
S.T.  4X₁ +  3X₂ +  2X₃ +  5X₄ ≤ 8
END
INT X₁
INT X₂
INT X₃
INT X₄
```

Then click on SOLVE. The output shows the optimal solution and the optimal value after eight Branch-and-Bound Iterations.

Note that instead of repeating INT four times, one can use INT 4. The first four variables appeared in the objective function.

```
        OBJECTIVE FUNCTION VALUE

        1)      24.00000

   VARIABLE           VALUE            REDUCED COST
       X₁            0.000000           -11.000000
       X₂            1.000000            -9.000000
       X₃            0.000000            -8.000000
       X₄            1.000000           -15.000000


       ROW   SLACK OR SURPLUSDUAL PRICES
        2)           0.000000              0.000000

  NO. OF ITERATIONS=  8
```

5.9.2 Example 5.8: Traveling Salesperson Problem

A salesperson has to visit cities 1, 2, ... n, and his/her trip begins at, and must end in, Home City. Let C_{ij} be the cost of traveling from city i to city j, which is given. The problem is to determine an optimal order for traveling the cities so that the total cost is minimized.

Consider the following Traveling Salesperson Problem:

Traveling Salesman Problem

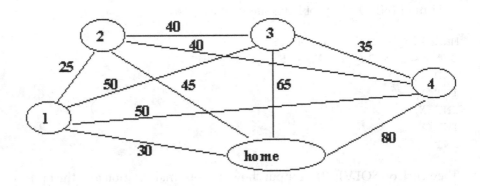

$$\min \quad \sum_i \sum_j d_{ij} X_{ij}$$

subject to $\quad \sum_j X_{ij} = 1$, for all i

$$\sum_i X_{ij} = 1, \text{ for all } j$$

$$X_{ij} = 0, \text{ or } 1$$

May need sub-tour breaker to be added.

$$\sum_i \sum_j X_{ij} \leq \quad \text{number of arc in the sub-tour-1}$$

The LP formulation is:

```
Min 30x01 + 45x02 + 65x03 + 80x04 + 25x12 + 50x13 + 50x14

+ 40x23 + 40x24 + 35x34 + 30x10 + 45x20 + 25x21 + 65x30
+ 50x31

+ 40x32 + 80x40 + 50x41 + 40x42 + 35x43
```

Subject to:

$$X_{01} + X_{02} + X_{03} + X_{04} = 1$$

$$X_{01} + X_{02} + X_{03} + X_{04} = 1$$

$$X_{10} + X_{12} + X_{13} + X_{14} = 1$$

$$X_{20} + X_{21} + X_{23} + X_{24} = 1$$

$$X_{30} + X_{31} + X_{32} + X_{34} = 1$$

$$X_{40} + X_{41} + X_{42} + X_{43} = 1$$

$$X_{10} + X_{20} + X_{30} + X_{40} = 1$$

$$X_{01} + X_{21} + X_{31} + X_{33} = 1$$

$$X_{02} + X_{12} + X_{32} + X_{42} = 1$$

$$X_{03} + X_{13} + X_{23} + X_{43} = 1$$

$$X_{04} + X_{14} + X_{24} + X_{34} = 1$$

All X_{ij} = 0 or 1

The solution to this LP problem produces a sub-tour (0, 1, 2). We need to introduce a tour breaker that is:

$$X_{01} + X_{10} + X_{12} + X_{21} + X_{02} + X_{20} \leq 2$$

Adding this new constraint and resolving, we need another tour breaker, which is:

$$X_{01} + X_{10} \leq 1,$$

Adding this, the optimal path is: Home to 1, 1 to 2, 2 to 4, 4 to 3, and 3 to home, with a total length of 195 units.

5.9.3 Example 5.9: Capital Budgeting Applications

Suppose a Research and Development company has a sum of money, D dollars, available for investment. The company has determined that there are N projects suitable for investment and at least d_j dollars must be invested in project j if it is decided that project j is worthy of investment. The company net profit that can be made by an investment in project j is P_j dollars. The company's dilemma is that it cannot invest in all N projects, because:

$\Sigma\ d_j > D$. Thus, the company must decide in which of the projects it wishes to invest in order to maximize its profit. To solve this problem, the counselor formulates the following problem:

Let $X_j = 1$, if the company invests in project j, and

$X_j = 0$ if the company does not invest in project j.

The total amount that will be invested is then

$\Sigma\ d_j\ X_j$, and since this amount cannot exceed D dollars, we have the constraint:

$$\Sigma\ d_j\ X_j \leq D$$

The total profit will be $\Sigma\ P_j\ X_j$. Thus, the company desires the 0–1 problem:

$$\text{Maximize } Z = \Sigma\ P_j\ X_j$$
$$\text{Subject to: } \Sigma\ d_j\ X_j \leq D, X_j = 0, \text{ or } 1.$$

5.9.4 Example 5.10: Marketing Application

Suppose five daily newspapers are being published in a certain country, each paper covering some of the nine regions of the country as shown in the following table:

Newspaper#	Region Covered	Cost of Advertisement	Benefit of Advertisement
1	1, 2, 3	3	12
2	2, 3, 6	4	10
3	4, 5, 6	3	14
4	5, 7, 8	7	19
5	6, 8, 9	5	16

The Marketing Manager problem is to find a minimum total cost such that the advertisement covers the whole country. This problem can be formulated as the following zero-one LP problem:

Minimize $C = 3y_1 + 4y_2 + 3y_3 + 7y_4 + 5y_5$

s.t. y_1		≥ 1	(Region 1)
$y_1 + y_2$		≥ 1	(Region 2)
$y_1 + y_2$		≥ 1	(Region 3)
y_3		≥ 1	(Region 4)
$y_3 + y_4$		≥ 1	(Region 5)
$y_2 + y_3 + y_5$		≥ 1	(Region 6)
y_4		≥ 1	(Region 7)
$y_4 + y_5$		≥ 1	(Region 8)
y_5		≥ 1	(Region 9)

$y_j = 0$ or 1, for all j's

The optimal solution is to advertise in Newspapers 1, 3, 4, and 5 for a total cost of $18.00. This solution is the lowest cost associated with coverage in each of the nine areas.

5.9.5 Example 5.11: The Cutting Stock Problem

Suppose that a lumberyard has a supply of 10-ft boards, which are cut into 3-ft, 4-ft, and 5-ft boards according to customer demand. The 10-ft boards can be cut in six different sensible patterns as shown in the following table:

Pattern #	# of 3-ft boards	# of 4-ft boards	# of 5-ft boards	Waste (ft)
1	3	0	0	1
2	2	1	0	0
3	1	0	1	2
4	0	1	1	1
5	0	2	0	2
6	0	0	2	0

There are many other possible but not sensible patterns; for instance, one would cut a 10-ft board into a 3-ft and a 4-ft board, leaving 3-ft as waste. This would not make sense, since the 3-ft waste could be used as a 3-ft board, as in pattern #2. If a customer orders 50, 3-ft boards and 65 4-ft boards, the question is how many 10-ft boards need to be cut and what cut pattern is to be used?

To model this decision problem, let's denote by y_j the number of 10-ft boards cut according to pattern j, $j = 1, \ldots, 6$. Whereas the total customer demand is $50(3) + 65(4) + 40(5) = 610$ ft, the total length of boards actually cut is:

$$10(y_1 + y_2 + y_3 + y_4 + y_5 + y_6),$$

and the total waste is, therefore:

$$10(y_1 + y_2 + y_3 + y_4 + y_5 + y_6) - 610$$

This implies that when we:

$$\text{minimize } y_1 + y_2 + y_3 + y_4 + y_5 + y_6,$$

the total number of 10-ft boards that need to be cut, we also minimize the total waste and vice versa. The actual number of 3-ft boards obtained in the cutting procedure is:

$$3y_1 + 2y_2 + y_3,$$

and therefore:

$$3y_1 + 2y_2 + y_3 \geq 50$$

must hold in order to satisfy customer demand for 3-ft boards. Similarly,

$$y_2 + y_4 + 2y_5 \geq 65$$

to satisfy the demand for 4-ft boards and

$$y_3 + y_4 + 2y_6 \geq 40$$

for 5-ft boards. Since the variables y_j must be nonnegative integers, $j = 1, \ldots, 6$, we can summarize this *cutting stock* formulation as:

P: Min $z = y_1 + y_2 + y_3 + y_4 + y_5 + y_6$ (10-ft boards)
s.t. $3\,y_1 + 2y_2 + y_3$ ≥ 50 (3-ft boards)
$y_2 + y_4 + 2y_5$ ≥ 65 (4-ft boards)
$y_3 + y_4 + 2y_6$ ≥ 40 (5-ft boards)
y_j are nonnegative integer, $j = 1, \ldots, 6$

The optimal solution to this problem, using your software package, is to cut a total of 65 10-ft boards, using pattern #2 25 times and pattern #5 and #6 20 times each. The total waste would then be $25 \times 0 + 20 \times 2 + 20 \times 0 = 40$ ft.

The above example illustrates a simple instance of the cutting stock or trim loss problem, which is widely applicable whenever trim waste is to be minimized. The example above refers to a one-dimensional situation (board length).

5.10 AN ENGINEERING APPLICATION: MIXING SUBSTANCES

A chemical company is producing two types of substances (A and B) consisting of three types of raw materials (I, II, and III). The requirements on the compositions of the three substances and the profits are as follows:

Substance	Compositions	Profits per kg
A	• At most 20% of I • At most 10% of II • At least 20% of III	10
B	• At most 40% of I • At most 50% of III	8

Each material's available amount and treatment costs are as follows:

Raw Material	Available Amount (kg)	Treatment Costs/kg
I	400	4
II	500	5
III	300	6

The company's problem is to find out how much of each substance to produce at which composition, such that the profit is maximized.

Because we have two substances and three raw materials, we introduce $6 (= 2 \times 3)$ decision variables x_{ij}, $i = $ A, B, and $j = $ I, II, III.

For example, x_{BIII} is the amount of raw material III in substance B. The goal function is thus to maximize the profits minus the treatment costs:

$$\text{Max } 10(x_{AI} + x_{AII} + x_{AIII}) + 8(x_{BI} + x_{BII} + x_{BIII}) - 4(x_{AI} + x_{BI}) - 5(x_{AII} + x_{BII}) - 6(x_{AIII} + x_{BIII})$$

Rewriting this in terms of the six decision variables, the equivalent problem is:

$$\max: 6x_{AI} + 5x_{AII} + 4x_{AIII} + 4x_{BI} + 3x_{BII} + 2x_{BIII}$$

The material constraints are:

$$x_{AI} + x_{BI} \leq 400$$

$$x_{AII} + x_{BII} \leq 500$$

$$x_{AIII} + x_{BIII} \leq 300.$$

The composition constraints are:

$$0.2(x_{AI} + x_{AII} + x_{AIII}) \geq x_{AI}$$

$$0.1(x_{AI} + x_{AII} + x_{AIII}) \geq x_{AII}$$

$$0.2(x_{AI} + x_{AII} + x_{AIII}) \leq x_{AIII}$$

$$0.4(x_{BI} + x_{BII} + x_{BIII}) \geq x_{BI}$$

$$0.5(x_{BI} + x_{BII} + x_{BIII}) \geq x_{BIII}$$

From the first inequality in the table above, we can derive the following constraint:

$$0 \leq -0.8x_{AI} + 0.2x_{AII} + 0.2x_{AIII}$$

Our problem has therefore five constraints from the table above plus three material constraints and one goal function. In addition, we require that the decision variables, x_{ij}, take on only nonnegative integer values.

Using our software package, the optimal solution is:

$$x_{AI} = 85,\ x_{AII} = 42,\ x_{AIII} = 299,\ x_{BI} = 306,\ x_{BII} = 458,\ x_{BIII} = 1,$$

with a total gain of 4,516.

This means that the company should produce 426 kg of substance A and 765 kg of substance B. Of the 400 kg of material I only 391 kg is used, while all of the 500 kg of material II and all of the 300 kg of material III are used.

5.11 EXERCISES

5.1. Consider the following manufacturing process, where we have three DMUs each of which has two inputs and three outputs as shown in the data table below. Using DEA, rank these DMUs.

DMU	Input#1	Input#2	Output#1	Output#2	Output#3
1	6	15	10	5	17
2	9	16	6	8	11
3	8	13	5	10	14

5.2. Consider the following manufacturing process, where we have three DMUs each of which has two inputs and three outputs as shown in the data table below. Using DEA, rank these DMUs.

DMU	Input#1	Input#2	Output#1	Output#2	Output#3
1	6.5	15	10	5.5	16.5
2	9	16.5	6	8.5	11
3	8	13	5.5	10	14

5.3. Given the following graph, compute the MST from a to g by each method.

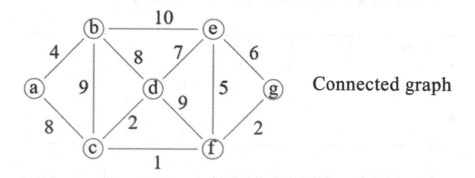

Connected graph

5.4. The Emergency Service Coordinator (ESC) for a county is interested in locating the county's three ambulances to maximize the residents that can be reached within 8 minutes in emergency situations. The county is divided into six zones, and the average time required to travel from

TABLE 5.2   Average Travel Times from Zone i to Zone j in Perfect Conditions

Zones	1	2	3	4	5	6
1	1	8	12	14	10	16
2	8	1	6	18	16	16
3	12	18	1.5	12	6	4
4	16	14	4	1	16	12
5	18	16	10	4	2	2
6	16	18	4	12	2	2

TABLE 5.3   Population in Each Zone

1	50,000
2	80,000
3	30,000
4	55,000
5	35,000
6	20,000
Total	270,000

one region to the next under semi-perfect conditions are summarized in Table 5.2.

The population in zones 1, 2, 3, 4, 5, and 6 is given by the Table 5.3.

Determine the location for the placement of the ambulances to maximize coverage within the allotted time.

5.5. Formulate and solve the following ax flow problem.

Source/Sink	1	2	3	4	5	6	7
1	—	10	10				
2			1	8		6	
3		1			12	1	
4						2	7
5						2	8

Apply the integer techniques in section 5,7 and 5,8 to solve the following problem.

5.6. Coach Bobby is trying to choose a starting lineup for his basketball team. His teach consists of seven players who have been rated on a scale of 1 = poor to 3 = excellent according to their abilities in their ball-handling, shooting rebounding, and overall defensive abilities. The positions that each player can play and their ability scores are in the table below. His five-player starting lineup must satisfy the following restrictions:

(1) At least four members must be able to play guard, at least two members must be able to play forward, and at least one member must be able to play center.

(2) The average ball-handling, shooting, and rebounding level of the starting lineup must be at least 2.

(3) If player 3 starts, then play 6 cannot start.

(4) If player 1 starts, then players 4 and 5 must both start.

(5) Either player 2 or player 3 must start.

The coach wants to maximize the defensive ability of the starting team.

Player	Position	Ball-Handling	Shooting	Rebounding	Defense
1	G	3	3	1	3
2	C	2	1	3	2
3	G-F	2	3	2	2
4	F-C	1	3	3	1
5	G-F	1	3	1	2
6	F-C	3	1	2	3
7	G-F	3	2	2	1

REFERENCES AND ADDITIONAL READINGS

Ahuja R., T. Magnanti, and J. Orlin, *Network Flows: Theory, Algorithms and Applications*, Prentice Hall, 1993.

Aliezhad, A., and A. Amini, Sensitivity analysis of topsis technique: The results of change in the weight of one attribute on the final ranking of alternatives, *Journal of Optimization in Industrial Engineering*, 7, 23–28, 2011.

Eiselt H., and C. Sandblom, *Integer Programming and Network Models*, Springer, 2000.

Floudas C., *Nonlinear and Mixed-Integer Optimization: Fundamentals and Applications*, Oxford University Press, 1995.

Neralic L., Sensitvity analysis in modeling data envelopment analysis, *Mathematical Communications*, 3(1), 1998.

Phillips N., *Network Models in Optimization & Their Applications in Practice*, Wiley, 1992.

Schrijver A., *Theory of Linear and Integer Programming*, Wiley, 1998.

Sierksma G., *Linear and Integer Programming: Theory and Practice*, Marcel Dekker, 2002.

Trick M. A., *Multiple Criteria Decision Making for Consultants*, 2012. http://mat.gsia.cmu.edu/classes/mstc/multiple.html (accessed April 2012).

Trick M. A., *Data Envelopment Analysis*, 2012. http://mat.gsia.cmu,edu/classes.QUANT/NOTES/chap12.pdf (accessed April 2012)

Winston W., *Introduction to Mathematical Programming*, Duxbury Press, pp. 322–325, 1995.

CHAPTER 6

Resource Allocation Models Using Dynamic Programming

I**N MANY REAL-WORLD PROBLEMS,** decisions must be made sequentially at different points in time or at different levels of components, subsystems, or major end items. A business' acquisition process is an example of a real-world sequential decision process. Since these decisions are made at several stages, they are often referred to as multistage decision problems. Dynamic programming (DP) is a mathematical technique that is well suited for this class of problems. Dynamic programming was developed by Richard Bellman in the early 1950s and is thus a relatively new methodology in the world of mathematics.

6.1 INTRODUCTION: BASIC CONCEPTS AND THEORY

Dynamic programming can solve both discrete and continuous nonlinear, multistage problems. Most texts present only the discrete DP algorithms. In this chapter, we will expose you to the continuous DP model, so we will begin with continuous DP and follow up with some discrete problems. The definitions and concepts that we define are valid in both discrete and continuous problems. There does not exist a standard mathematical formulation for any DP problem. DP is a general approach to solving problems that need to be solved in stages. Therefore, a certain amount of ingenuity and creativity is required to formulate and solve a

DOI: 10.1201/9781003464969-6

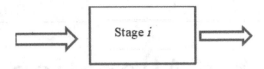

FIGURE 6.1   Typical Stage in a DP Problem.

DP problem. Exposure to examples will increase your ability to formulate and solve DP problems.

Let's begin with a more general framework and an easier example.

If we are going to consider DP as a multistage decision process, then let's closely examine one of the stages. Figure 6.1 illustrates a single-stage decision problem. A decision process of this type is characterized by Inputs, Decision Variables, Outputs (for the next stage), and the Return function, which measures the effectiveness of the current decisions. Multistage problems are a series of these types of stages that are linked together. The Inputs to subsequent stages are the Outputs of the previous stage.

There exist three general types of DP problems. They are the initial value, the final value, and the boundary value problems.

In the initial value problems, the initial state variable is prescribed. In the final value problems, we know the value of the final state variable, and in the boundary value problems, we know both the initial and the final state variable values. DP is a recursive algorithm, and, in many cases, it is best to work, systematically, backwards to solve the problem. This makes use of the concept of suboptimization and the principle of optimality. Bellman (1957) defined his principle of optimality as follows:

> An optimal policy (or set of decisions) has the property that whatever the initial state and initial decisions are, the remaining decisions must constitute an optimal policy with regard to the state resulting from the first decision.

This means that for every possible decision that could have been made from the first stage, all subsequent decisions are optimal with regard to the previous-stage decision. This will become clear as we do an example.

Before we begin an example, we should examine the recurrence relationships. Assume the desired objective is to maximize an n-stage objective function, f, which is given as the sum of all individual stage returns:

$$\text{Maximize } f = R_n\left(x_n, s_{n+1}\right) + R_{n-1}\left(x_{n-1}, s_n\right) + \cdots + R_1\left(x_1, s_2\right)$$

where the state and decision variables are related by

$$s_i = t_i\left(s_{i+1}, x_i\right) \quad i = 1, 2, \ldots, n$$

By the definition of the principle of optimality and the recursive property, we start at the final stage with input as specified and find x_i to optimize the return. Irrespective of what happens in the other stages, x_i must be found such that R is optimized for the given inputs.

Let's reexamine the definition and characteristics of a DP problem. DP is a technique that can be used to solve many optimization problems. In most applications, DP obtains solutions by working backwards from the end of the problem toward the beginning. Thus, it breaks up a long problem into a series of smaller, tractable problems.

6.2 CHARACTERISTICS OF DYNAMIC PROGRAMMING

The following dynamic programming characteristics are provided:

- The problem must be divisible into stages with a decision at each stage.

- Each stage has a number of states associated with it. By state, we mean the information that is needed at any stage to make an optimal solution.

- Often, the optimal solution will be in a variable form (temporarily).

- The decision chosen at any stage describes how the state at the current stage is transformed into the state at the next stage.

If the states have been classified into one of T stages, there must be a recursion that reflects the cost or reward earned from stages, $t, t+1, t+2, \ldots, T$ to the cost and reward earned from stages $t+1, t+2, \ldots, T$. In essence, this is a backward-working procedure. If the number of stages is sufficiently large, it can create computational difficulties.

6.2.1 Working Backwards

The stage is the mechanism by which we build up the problem. The state at any stage gives the information needed to make the correct decision at the current state. In most cases, we must determine how the reward (cost) during the current stage depends upon the stage decision, states, and value of t.

We must determine how the stage $t + 1$ depends upon the stage t decision.

If we define (minimization problem) ft (i) as the minimum cost incurred in stages $t, t + 1, t + 2, \ldots, T$, given that the stage t state is i, then (in many cases) we may write,

ft(i) = MIN{(Cost during Stage t) + ft + 1(new state at stage t +1)},

where minimization is over all possible decisions at that stage.

We begin the process by determining all the fT(.)s, then all the fT–1(.)'s, back to f1(initial).

We then determine the optimal stage 1 decision. This leads to stage 2 where we find the optimal stage 2 decision. We continue until stage T is found.

Let us illustrate discrete DP through two discrete (integer) resource allocation problems.

6.2.2 Example 6.1: A Knapsack Problem

Consider a knapsack that can hold 8 pounds. Item 1 weighs 4 lbs, item 2 weighs 3 lbs, and item 3 weighs 2 lbs. The returns (gain) for placing these items in the knapsack are 5, 4, and 2, respectively. How many of each item should we place in the knapsack to maximize the return?

$$\text{Max } w = 5x + 4y + 2z$$
$$\text{s.t. } 4x + 3y + 2z \le 8$$
$$x, y, z \ge 0, \text{ and all integer}$$

We draw a three-stage diagram, as given in Figure 6.2

The transitions are as follows:

Stage 3:

$$S_3 = 8$$

Stage 2:

$$S_2 = S_3 - 2z$$

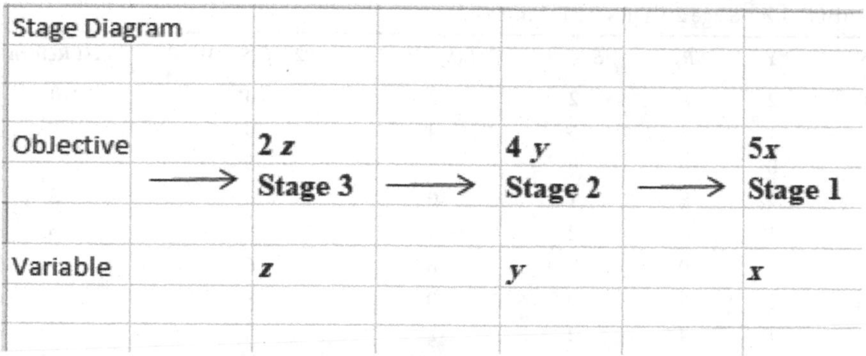

Stage Diagram				
Objective	$2z$ $\longrightarrow$ Stage 3	$\longrightarrow$	$4y$ Stage 2	$\longrightarrow$	$5x$ Stage 1
Variable	z		y	x

FIGURE 6.2   Stage Diagram for Example 6.1.

TABLE 6.1   Stage 3 Values for Example 6.1

S_3	z	R_3
8	4	8
7	3	6
6	3	6
5	2	4
4	2	4
3	1	2
2	1	2
1	0	0
0	0	0

Stage 1:

$$S_1 = S_2 - 3y$$

Solution is:
Stage 3 (Table 6.1):

Objective Function: Maximize $R_3 = 2z$

Stage 3 (Table 6.2):

Objective function: Maximize $R_2 + f_3(S_3)$

TABLE 6.2   Stage 2 Values for Example 6.1

S_2	Y	R_2	$S_2 - 3y$	$f_3(S_2 - 3y)$	$R2 + f_3(S_2 - 3y)$	Best Return
8	2	8	2	2	10*	10
	1	4	5	4	8	
	0	0	8	8	8	
7	2	8	1	0	8*	8
	1	4	4	4	8*	8
	0	0	7	6	6	
6	2	8	0	0	8*	8
	1	4	2	2	6	
	0	0	6	6	6	
5	1	4	2	2	6*	6
	0	0	5	1	2	
4	1	4	1	0	4*	4
	0	0	4	1	2	
3	1	4	0	0	4*	4
	0	0	3	1	2	
2	0	0	2	1	2*	2
1	0	0	1	0	0	0
0	0	0	0	0	0	0

Stage 1 (Table 6.3):

TABLE 6.3   Stage 1 Values for Example 6.1

$S1$	X	$R1$	Best f_2 $(S_1 - 4x)$	$f_1 = R_1 + $ Best f_2
8	2	10	0	10
	1	5	4	9
	0	0	10	10

The optimal objective function value is 10. This is achieved by multiple optimal solutions as evidenced by the value 10 turning up as the best result in several final columns. Let's backtrack to find all the multiple optimums. When $x = 2$, nothing is based on the remaining stages, so $y = z = 0$. When $x = 0$, all 8 lbs are remaining for y and z to share. In Stage 2, with 8 lbs to use, the best solution was 10 with $y = 2$ and $z = 1$ (Table 6.4).

TABLE 6.4   Optimal Values for Example 6.1

X	Y	Z	W
0	2	1	10
2	0	0	10

6.3 MODELING AND APPLICATIONS OF DISCRETE DYNAMIC PROGRAMMING

We will start this section with a problem looking at the number of drug-related incidents in high schools. The number of drug-related incidents in each of a city's three high schools depends upon the number of security patrolpersons assigned to each school. The city has five security patrolpersons available for assignment to the three schools. Historical records show the number of incidents due to suspected drugs as given in Table 6.5.

Determine how to assign patrolpersons to minimize the number of drug-related incidents.

We find that our optimal minimum solution is 37 events. There are alternate solutions that can achieve this result.

6.3.1 Oil Well Investment DP Application

We have $4 million to invest in four oil wells. The amount of revenue earned at each of the four sites depends on the amount of investment in each site. This information is provided in the table below and was derived from revenue forecasting formulas. If the revenue invested in each site must be multiples of $1 million, use DP to determine the optimal investment policy to maximize revenues (Table 6.6).

TABLE 6.5   School Drug-Related Incidents.

		No. of Patrol Persons Assigned to Schools				
		0	1	2	3	4
School 1	14	10	7	4	1	0
School 2	25	19	16	14	12	11
School 3	20	14	11	8	6	5

TABLE 6.6   Expected Roil Revenue

Amount Invested ($Millions)	Revenue ($Millions)			
	SITE 1	SITE 2	SITE 3	SITE 4
0	4	3	3	2
1	7	6	7	4
2	8	10	8	9
3	9	12	13	13
4	11	14	15	14

Solution Process:

(I). SITE 1 (Table 6.7):

TABLE 6.7   Site 1 Solution

S_1	D_1	R_1
4	4	11
3	3	9
2	2	8
1	1	7
0	0	4

(II) SITE 2 (Table 6.8):

TABLE 6.8   Site 2 Solution

S_2	D_2	R_2	S_1	$f^*(S_1)$	Sum	Best
4	4	14	0	4	18	
	3	12	1	7	19	19
	2	10	2	8	18	
	1	6	3	9	15	
	0	3	4	11	14	
3	3	12	0	4	16	
	2	10	1	7	17	17
	1	6	2	8	14	
	0	3	3	9	12	
2	2	10	0	4	14	14
	1	6	1	7	13	
	0	3	2	8	11	
1	1	6	0	4	10	10
	0	3	1	7	10	10
0	0	4	0	3	7	7

(III). SITE 3 (Table 6.9):

TABLE 6.9   Site 3 Solution

S_3	D_3	R_3	S_2	$f^*(S_2)$	Sum	Best
4	4	15	0	7	22	
	3	13	1	10	23	
	2	8	2	4	22	

TABLE 6.9 *(Continued)* Site 3 Solution

S_3	D_3	R_3	S_2	$f^*(S_2)$	Sum	Best
	1	7	3	17	24	24
	0	3	4	19	22	
3	3	13	0	7	20	
	2	8	1	10	18	
	1	7	2	14	21	21
	0	3	3	17	20	
2	2	8	0	7	15	
	1	7	1	10	17	17
	0	3	2	14	17	17
1	1	7	0	7	14	14
	0	3	1	10	13	
0	0	3	0	7	10	10

(IV) SITE 4 (Table 6.10):

TABLE 6.10 Site 4 Solution

S_4	D_4	R_4	S_3	$f^*(S_3)$	Sum	Best
4	4	14	0	10	24	
	3	13	1	14	27	27
	2	9	2	17	26	
	1	4	3	21	25	
	0	2	4	24	26	

Optimal solution is $27 million in revenues when we invest as follows:

SITE 4: $3 million

SITE 3: $1 million

SITE 2: 0

SITE 1: 0

6.4 EXERCISES

6.1. We have three projects competing for budget allocation. Let D_j equal the allocation of the budget to project j (j = 1, 2, 3). Let $r_j(D_j)$

TABLE 6.11   Expected Return for Exercise 6.1

D_j	$r_1(D_1)$	$r_2(D_2)$	$r_3(D_3)$
0	0	2	0
1	4	6	5
2	7	8	9
3	9	10	11
4	12	11	10

equal the return from project j for the given input D_j. In general, we desire to:

$$\text{Max } \Sigma r_j\left(D_j\right)$$
$$\text{s.t. } \Sigma D_j \leq K \quad (\text{Budget})$$
$$D_j \geq 0$$

Table 6.11 shows the return for each project type for the finite inputs D_j.

Our budget, K, is 4 units. Find the optimal D_j (for $j = 1, 2, 3$) to allocate from the budget using DP. What is the projected return for these projects?

6.2. Use DP to solve:

$$\text{Minimize } x^2 + y^2 + z^2$$
$$\text{s.t. } x + 2y + z \geq 9$$
$$x, y, z \geq 0 \ \& \ \text{all integer}$$

6.3 Use DP to solve:

$$\text{Minimize } x^2 + y^2 + w^2 + z^2$$
$$\text{subject to}: x + 2y + 2w + z \geq 9$$
$$x, y, w, z \geq 0, \text{ integer}$$

REFERENCES AND SUGGESTED READINGS

Bellman, R. *Dynamic Programming*. Princeton, NJ: Princeton University Press, 1957.

Bradley, S., A. Hax, and T. Magnanti. *Applied Mathematical Programming*. Reading, MA: Addison-Wesley, 1977.

Hillier, F., and G. Lieberman. *Introduction to Mathematical Programming*. New York: McGraw-Hill, 1990.

Luenberger, D. *Linear and Nonlinear Programming*. Reading, MA: Addison-Wesley, 1984.

Phillips, D., A. Ravindran, and J. Solberg. *Operations Research*. New York: John-Wiley & Sons, 1976.

Rao, S. *Optimization Theory and Applications*. New Delhi: Wiley Eastern Ltd., 1979.

Winston, W. *Introduction to Mathematical Programming: Applications and Algorithms*. Boston, MA: PWS-Kent, 1991.

CHAPTER 7

Queuing Models

QUEUING THEORY IS FOCUSED on understanding how lines form, how they function, and why they do not always work as expected. It examines every component of waiting in line to include the arrival process, time in system, and the number of customers, which might be people, data packets, cars, or anything else. Applications of queuing cover a wide range of businesses and attempts to provide faster customer service, increase traffic flow, improve order shipments, and improve wait times at data networks and call centers.

7.1 INTRODUCTION TO QUEUING THEORY

Have you ever driven up to a fast-food establishment and seen cars going around the block for the drive-through? There are models that address the rate at which cars arrive, orders are placed, and orders served. These options, to some degree, occur in banks, grocery stores, and other commercial retailers.

We will begin this section with a typical customer service problem example that is updated and modified from Fox and Burks (2022).

7.1.1 Example 7.1: Simple Fast-Food Service Queue

Let us think about a queue (waiting line) most of us have seen, the line at any store, fast-food chains, or bank. Let us consider the simplest case, one line in front of a single customer service cashier or automatic order machine. For now, ignore the role of second cashier (occasionally open) and the others. We will present later the wide applicability (not just cafes) of the insights that we can draw from this model.

 DOI: 10.1201/9781003464969-7

7.1.1.1 Inputs

Customers will arrive at the fast-food restaurant and are served immediately or join the line. The time between any two customer arrivals is variable and uncertain, but we can assume for a moment that on average two customers join the line every minute. The rate at which customers join the line is what we call arrival rate (symbol lambda λ).

In this case, arrival rate lambda λ is 2 per minute. This also means that the average time between two arrivals is ½ minute; we call this interarrival time. Let's assume that this is exponential.

The time the cashier takes to serve one customer is called the service time. Service time will change from customer to customer and is, therefore, variable and uncertain. We will assume that on average, the service time for a customer is 20 seconds = 1/3 minute. This also means that one cashier can serve customers at the rate of 1/(1/3) = 3 customers per minute. This is known as the service rate (symbol mu or μ). Let's assume that the service rate distribution is exponential.

Since there is only one cashier, the number of cashiers is one. We use the symbol m for the number of servers. If the restaurant opens another cashier and if a *single line* is used to feed both servers, the number of servers m is 2. In case there are two different lines in front of two cashiers then we will say that there are two different queues, each with $m = 1$. For now, we will stick with our original scenario of single queue with single server.

To summarize, at least three inputs are needed to define a queue: Arrival rate (λ), service rate (μ), and number of servers (m). Note that we are interested in expressing our inputs n terms of *rates* (e.g., per minute) and not in time (e.g., minutes).

Depending on the situation, we may need other inputs describing the extent of variability in arrivals and service. For this basic model, we assume a certain type of variability (Poisson distribution for the number of arrivals with interarrival times being exponential and an exponential distribution for service times) and not worry about it for now.

7.1.1.2 Utilization

The first output we can get is the utilization of our resource, the cashier. We use the symbol rho, ρ, for the utilization. This is equal to the ratio of the rate at which work arrives and the capacity of the server (cashier or teller for a bank). We know that the rate at which work arrives is arrival rate lambda, λ. One server can service the work at the rate of service rate μ. If there are more than one server (i.e., if number of servers m is more than

1), then *total* rate at which work can be served, station capacity, will be m multiplied by μ. Therefore, utilization rho ρ can be calculated as lambda λ divided by (m multiplied by μ). $\rho = \dfrac{\lambda}{m\mu}$.

Make sure arrival rate and service rate are expressed in the same unit of time; for example, both should be per minute or per hour. In the case of our establishment,

$$\rho = 2/3 = 0.66666$$

Our cashier/teller is busy about 67% of the time. For our calculations to work, *utilization rho ρ must be less than 1.*

7.1.1.3 Number of Customers Waiting

The second output is the number of customers waiting. We use the symbol L_q for this. Note that this *excludes* the person who is getting serviced by the cashier. We have a simple table to find this value. The table is attached. To look at the table, we need two things: first, the arrival rate/service rate, (i.e., lambda/mu, λ / μ) and second, number of servers m.

In our establishment case, λ / μ is $2/3 = 0.6667$ and m is 1.

You will find one row in the table that corresponds to these two values. In this row, read the number in column titled L_q. You will see that the closest number is 1.207. This means that, on average, the number of people waiting in line L_q is 1.207.

Only in the special case when there is only one server, $m = 1$, we can also use a simple formula to compute the number of customers waiting, L_q. (The table works for all values of m, including $m = 1$). For $m = 1$: $L_q = \dfrac{\rho^2}{1-\rho}$. For example, in this case,

$$L_q = \frac{(2/3)^2}{\left(1-\left(\dfrac{2}{3}\right)\right)} = 1.3333$$

which is not the same as we got from table because $\lambda / \mu = 0.6667$ is not a value in the table.

7.1.1.4 Waiting Times

The third output is the time an average customer waits in line to receive service. We use the symbol W_q for this. We have a simple formula to convert number-of-customers-waiting L_q into time-a-customer-waits W_q. To get W_q, divide L_q by arrival rate lambda λ (Little's law). In our establishment, the time-a-customer-waits is:

$$W_q = \frac{L_q}{\lambda} = \frac{1.33333}{2} = 0.6667 \text{ min } or \text{ about 40 secounds}$$

Sometimes, we want to think about the whole system, that is, not just waiting but both waiting and getting service. We would like to know the time-a-customer-spends-in-the-system (we use symbol W_s for this) including both the time for waiting and time for service. Clearly, this is equal to time-a-customer-waits W_q plus service time. In our example case, W_s is just equal to the sum of waiting time (40 sec) and service time (20 sec).

$$W_s = 60 \text{ sec} = 1 \text{ min}$$

7.1.1.5 Number-of-Customers-in-System

There is also the question of the number-of-customers-in-system (symbol L_s), including both the customers who are waiting and who are getting service. Another application of Little's law shows that to get number-of-customers-in-system symbol L_s, multiply time-in-system W_s by arrival rate lambda λ. In our case,

$$L_s = \lambda \, W_s = 2 \times 1 = 2$$

7.1.1.6 Idle Percentage

Finally, to compute the chance that system is idle, that is, there is no customer in the system, we can read the closed column value titled P_0 from the table, just the way we read L_q.

For this model, $P_0 = 0.350$, that is, there is 35% chance that the cashier is free (idle). We are often concerned with idle and busy times of the server.

7.1.1.7 Other Performance Measures

- For a single-server case, some other performance measures can be computed as following:

- Probability that there are n customers in system $P_n = (1 - \rho)\rho^n$

- Probability that the wait is greater than $t = \rho \, e^{-\mu(1-\rho)t}$

- Probability that time-in-system is greater than $t = e^{-\mu(1-\rho)t}$
- For more than one server, spreadsheets are available to compute these measures.

7.1.1.8 Determining Capacity

If we increase capacity (by increasing m or by increasing μ), we expect that the cost of providing that capacity will increase. We also expect, however, that the customers will wait less and that the cost of customer waiting will decrease. This suggests that we should look at the total cost = (cost of providing service + cost of customer waiting) in order to make a decision about how much capacity to provide.

For example, if our establishment pays $20 per hour to a cashier, then adding one more cashier/teller increases the cost of providing capacity by $20 per hour. If we add another server, (m goes from 1 to 2), then $\lambda / \mu = 2/(2 \times 3) = 1/3$, $L_q = 0.16667$, $W_q = 0.16667/2 = 0.08333$, $W_s = 5\ sec + 20\ sec = 25\ sec = ¼\ min$ and $L_s = 0.5$.

It also reduces the number of customers in the system from $L_s = 2$ (from above) to $L_s = 0.5$. If we assume that a customer's time is worth by paying $25 per hour, then the system saves $(1 - 0.5) \times \$25$ per hour = $12.50 which is less than the $20 that we pay the second server. Therefore, in this example, from the total system cost perspective, we should not add another server.

7.1.1.9 Other Modeling Extensions

We have made two significant assumptions about the pattern of variability in arrivals and service: Poisson distribution for number of arrivals and Exponential distribution for service times. These assumptions mean the following: Coefficient of variation = (Standard deviation/mean) for inter-arrival times, $C_a = 1$ and coefficient of variation = (Standard deviation/mean) for service times, $C_s = 1$. But what if based on measurement of real data, they are not 1? We call this the case of general arrivals and service.

It is easy to compute L_q in this more general case as follows:

$$L_q \text{ in case } C_a \neq 1 \text{ and / or } C_s \neq 1$$
$$= (L_q \text{ as computed from table}) \left(\frac{C_a^2 + C_s^2}{2} \right)$$

Starting from L_q, other performance measures can be computed in the same way as earlier.

7.1.1.10 Summary of Variables and Formulas

- Arrival rate lambda, $\lambda = 1/$(interarrival time, that is time between two arrivals)

- Service rate mu, $\mu = 1/$service time

- Number of servers m

- Utilization rho, $\rho = \lambda / (m\mu)$

- Assume arrivals Poisson distribution and service time are exponentially distributed.

- Average number in waiting line L_q can be obtained from table (given λ / μ and m)

- In case number of servers $m = 1$, we can also use $L_q = \rho^2 / (1 - \rho)$

- Average waiting time $W_q = L_q / \lambda$ from *Little's law*

- Average time-in-system (waiting time + service time) $W_s = W_q + (1/\mu)$

- Average number-in-system (waiting + getting served) $L_s = \lambda W_s$

- Probability that there is nobody in the system P_0 is available in table.

For single-server case, $m = 1$, we have following three formulas:

- Probability that there are n customers in system $P_n = (1 - \rho)\rho^n$

- Probability that wait is greater than t $= \rho e^{-\mu(1-\rho)t}$

- Probability that time-in-system is greater than t $= e^{-\mu(1-\rho)t}$

Determination of capacity is a trade-off between cost-of-service capacity and cost of customer waiting.

- Coefficient of variation of interarrival times C_a = (Std. dev./Mean of interarrival times)

- Coefficient of variation of service times, C_s = (Std. dev./Mean of service times)

If $C_a \neq 1$ *and/or* $C_s \neq 1$, modify L_q from above by multiplying it with $(C_a^2 + C_s^2)/2$

7.2 THE MULTI-SERVER PROBLEMS

The typical shorthand notation for service (queuing) models is $m/m/\#$. The first m refers to the assumed distribution on time between customer arrivals, the second m is the assumed distribution on service times, and the number provides the number of servers in the system. In the last section, we examined the $m/m/1$, so it is only right to look at the next logical extension of this model, the $m/m/2$ queue. This means that our system will have two servers.

7.2.1 Example 7.2: Local Hospital Service $m/m/2$

Consider that a small town, with one hospital, has two ambulances to supply ambulance service. Requests for ambulances during non-holiday weekend averages 0.8 per hour and tend to be Poisson distributed. Travel and assistance time averages one hour per call and follows an exponential distribution. What is the utilization of ambulances? On average, how many requests are waiting for ambulances? How long will a request have to wait for ambulances? What is the probability that both ambulances are sitting idle at a given point in time? Do any of these measures indicate a need for additional ambulances?

Ambulances are resources or servers servicing the requests that are coming in (customers). Two ambulances mean $m = 2$. Requests arrival rate is $\lambda = 0.8$ per hour. Service time = 1 hour.

Service rate: $\mu = (1/\text{service time}) = 1$ per hour.

Utilization: $\rho = \dfrac{\lambda}{m\mu} = \dfrac{0.8}{2*1} = 0.4$

$\lambda / \mu = 0.8$ and $m = 2$.

For $m = 2$, the formulas for L_q is $\dfrac{\lambda^3}{\mu\left(4\mu^2 - \lambda^2\right)}$ and for $P_o = \dfrac{1-\rho}{1+\rho}$

and for $L_q = 0.152$, $P_0 = 0.429 = (1 - .4)/(1 + .4) = 0.042857$

On an average, these many requests are waiting for ambulances = $L_q = 0.152$

A request will have to wait for ambulances for time $W_q = \dfrac{L_q}{\lambda} = \dfrac{0.152}{0.8} = 0.19 \, \text{hr}$

The probability that both ambulances are sitting idle at a given point in time Is: $P_0 = 0.429$

Other useful performance measures include W_s, L_s.

Time spent by a call in system.

$$W_s = W_q + \text{service time} = 0.19 + 1 = 1.19 \text{hr}$$

Number of requests in system $= L_s = \lambda W_s = 0.8 * 1.19 = 0.952$

Our determination here is that two ambulances are sufficient.

7.2.2 Example 7.3: Local Car for Hire, m/m/2 Queue Revisited

Let's revisit our small town. It currently has three local cars for hire to pick up passengers for short trips around town. Requests for service on Friday night are an average of 4 per hour and tend to be Poisson distributed. Travel time averages 15 minutes per pickup and follows an exponential distribution. What is the utilization of cars? On average, how many customers are waiting for pickup? How long will a customer have to wait for a car? What is the probability that all cars are sitting idle at a given point in time? Do any of these measures indicate a need for additional cars?

Ambulances are resources or servers servicing the requests that are coming in (customers). Three cars mean $m = 2$. Requests' arrival rate $\lambda = 4$ per hour. Service time $= 0.25$ hour.

Service rate $\mu = (1/\text{service time}) = 4$ per hour.

Utilization: $\rho = \dfrac{\lambda}{m\mu} = \dfrac{4}{2 * 4} = 0.5$

$\lambda / \mu = 1$ and $m = 2$.

For $m = 2$, the formula for L_q is $\dfrac{\lambda^3}{\mu(4\mu^2 - \lambda^2)}$ and for Po it is $\dfrac{1-\rho}{1+\rho}$

and for $L_q = 0.333$, $P_0 = 0.333 = (1 - 0.5)/(1 + 0.5) = 0.333$

On an average, these many requests are waiting for a car $= L_q = 0.333$.

A request will have to wait for a car for $W_q = \dfrac{L_q}{\lambda} = \dfrac{0.333}{4} = 0.08 \, hr.$

The probability that both cars are sitting idle at a given point in time is $P_0 = 0.333$.

Other useful performance measures include W_s, L_s.

Time spent by a call in system is:

$$W_s = W_q + \text{service time} = 0.08 + 0.25 = 0.33\,\text{hr}$$

Number of requests in system $= L_s = \lambda W_s = 4 * 0.33 = 1.32$

What we find here is that two cars are not sufficient to meet demands.

7.3 EXERCISES

7.1. A small town with one hospital has two ambulances to supply ambulance service. Requests for ambulances during non-holiday weekend averages 0.75 per hour and tend to be Poisson distributed. Travel and assistance time averages 1 hour per call and follows an exponential distribution. What is the utilization of ambulances? On an average, how many requests are waiting for ambulances? How long will a request have to wait for ambulances? What is the probability that both ambulances are sitting idle at a given point in time?

7.2. At a bank's ATM location with a single machine, customers arrive at the rate of one every other minute. This can be modeled using a Poisson distribution. Each customer spends an average of 90 seconds completing his/her transactions. Transaction time is exponentially distributed. Determine: (1) The average time customers spend from arriving to leaving, (2) the chance that the customer will not have to wait, and (3) the average number waiting to use the machine.

7.3. The last two things that are done before a car is completed are engine marriage (station 1) and tire installation (station 2). On an average, 54 cars per hour arrive at the beginning of these two stations. Three servers are available for engine marriage. Engine marriage requires 3 minutes. The next stage is a single-server tire installation. Tire installation requires 1 minute. Arrivals are Poisson distributed, and service times are exponentially distributed.

(1) What is the queue length at each station?

(2) How long does a car spend waiting at the final two stations?

7.4. A machine shop leases grinders for sharpening their machine cutting tools. A decision must be made as to how many grinders to lease. The cost to lease a grinder is $50 per day. The grinding time required by a machine operator to sharpen his cutting tool has an exponential distribution, with an average of 1 minute. The machine operators arrive to sharpen their tools according to a Poisson process at a mean rate of one every 20 seconds. The estimated cost of an operator being away from his machine to the grinder is 10¢ per minute. The machine shop is open 8 hours per day. How many grinders should the machine shop lease?

7.5. Consider a queue with a single server, arrival rate of 5 per hour, and service rate of 10 per hour. Assuming Poisson arrivals and exponential service time, what is the waiting time in queue?

(a) Actual measurements show that interarrival time standard deviation is 24 minutes, and service time standard deviation is 3 minutes. What is the waiting time in queue?

REFERENCES AND SUGGESTED READINGS

Fox, W., Burks, R. (2022). *Modeling Change and Uncertainty: Machine Learning and Other Techniques*. CRC Press, Boca Raton, FL.

Giordano, F., Fox, W., Horton, S. (2013). *A First Course in Mathematical Modeling*. 5th Ed. Boston, MA: Cengage Publishers.

Law, A., Kelton, D. (2007). *Simulation Modeling and Analysis*. 4th Ed. New York: McGraw Hill.

Meerschaert, M. M. (1993). *Mathematical Modeling*. San Diego: Academic Press.

Winston, W. (1994). *Operations Research: Applications and Algorithms*. 3rd Ed. Belmont, CA: Duxbury Press.

CHAPTER 8

Simulation Models

A MODELER MAY ENCOUNTER SITUATIONS where the construction of an analytic model is infeasible because of the complexity of the situation. In instances where the behavior cannot be modeled analytically or where data are collected directly, the modeler might simulate the behavior indirectly and then test various alternatives to estimate how each affects the behavior. Data can then be collected to determine which alternative is best. Monte Carlo simulation is a common simulation method that a modeler can use, usually with the aid of a computer. The proliferation of today's computers in the academic and business worlds makes Monte Carlo simulation very attractive. It is imperative that students have at least a basic understanding of how to use and interpret Monte Carlo simulations as a modeling tool.

Simulation modeling is the process of creating prototype model to predict and analyze its performance in the real world. For example, engineers use simulation modeling to understand the conditions in which ways a part could fail or what loads it can withstand. Simulation models enable us to employ a stochastic approach to solving problems that do not have a closed form solution. Monte Carlo Simulation provides a powerful simulation approach to help us in cases where we cannot find a closed form solution.

In this chapter, we present algorithms and Excel output for the following simulation applications.

1. An aircraft missile attack

2. The amount of gas that a series of gas stations will need

3. A simple single barber in a barbershop queue

DOI: 10.1201/9781003464969-8

8.1 MISSILE ATTACK

An analyst plans a missile strike using F-15 aircraft. The F-15 must fly through air-defense sites that hold a maximum of eight missiles. It is vital to ensure success early in the attack. Each aircraft has a probability of 0.5 of destroying the target, assuming it can get to the target through the air-defense systems and then acquire and attack its target. The probability that a single F-15 will acquire a target is approximately 0.9. The target is protected by air-defense equipment with a 0.30 probability of stopping the F-15 from either arriving at or acquiring the target. How many F-15 are needed to have a successful mission, assuming we need a 99% success rate?

8.1.1 Algorithm: Missiles

Inputs: N = number of F-15s

M = Number of missiles fired

P = Probability that one F-15 can destroy the target

Q = Probability that air defense can disable an F-15

Output: S = Probability of mission success

Step 1.   Initialize $S = 0$

Step 2.   For $I = 0$ to M do

Step 3. $P(i) = [1 - (1 - P)^{N-I}]$

Step 4. $B(i)$ = Binomial distribution for (m, i, q)

Step 5.   Compute $S = S + P(i) * B(i)$

Step 6.   Output S.

Step 7.   Stop

We run the simulation (Figure 8.1) letting the number of F-15s vary and calculate the probability of success. We guess $N = 15$ and find that there is a probability of success greater than 0.99 when we send nine planes. Thus, any number greater than 9 works.

	S	Initial S		Bombers	N	p	0.5	T	0.9	P*T	0
18											
19			Initial S		Bombers	N	q	0.3				
20	S	0			15	Quess				S > 99	good
21									S_Final	0.99313666	
22	i	B	P	P*B	New S						
23	0	0.004747562	0.9999	0.004747	0.004747						
24	1	0.030520038	0.9998	0.030513	0.03526						
25	2	0.091560115	0.9996	0.091522	0.126781						
26	3	0.170040213	0.9992	0.16991	0.296691						
27	4	0.218623131	0.9986	0.218319	0.51501						
28	5	0.206130381	0.9975	0.205608	0.720618						
29	6	0.147235986	0.9954	0.146558	0.867176						
30	7	0.081130033	0.9916	0.080451	0.947627						
31	8	0.034770014	0.9848	0.034241	0.981867						
32	9	0.011590005	0.9723	0.011269	0.993137						
33	10	0.002980287	0.9497	0.00283	0.995967						
34	11	0.000580575	0.9085	0.000527	0.996494						
35	12	8.29393E-05	0.8336	6.91E-05	0.996564						
36	13	8.20279E-06	0.6975	5.72E-06	0.996569						
37	14	5.02212E-07	0.45	2.26E-07	0.996569						
38	15	1.43489E-08	0	0	0.996569						

FIGURE 8.1 Excel Screenshot of Missile Attack.

We find that nine F-15s give us $P(s) = 0.99313$.

Actually, any number of F-15 greater than 9 provides a result with the probability of success we desire. Fifteen F-15s yield a $P(s) = 0.996569$. Any more would be overkill.

8.2 GASOLINE-INVENTORY SIMULATION

You are a consultant to an owner of a chain of gasoline stations along a freeway. The owner wants to maximize profits and meet consumer demand for gasoline. You decide to look at the following problem.

8.2.1 Problem

Minimize the average daily cost of delivering and storing sufficient gasoline at each station to meet consumer demand.

8.2.2 Assumptions

For an initial model, consider that, in the short run, the average daily cost is a function of demand rate, storage costs, and delivery costs. You also assume that you need a model for the demand rate. You decide that historical data will assist you (Giordano et al., 2014). This information is displayed in Tables 8.1 and 8.2.

Model: We convert the number of days into probabilities by dividing by the total and we use the midpoint of the interval of demand for simplification.

TABLE 8.1   Gasoline Demand

Demand (Gallons)	Number of Occurrences (Days)
1,000–1,099	10
1,100–1,199	20
1,200–1,299	50
1,300–1,399	120
1,400–1,499	200
1,500–1,599	270
1,600–1,699	180
1,700–1,799	80
1,800–1,899	40
1,900–1,999	30
Total Days =	1,000

TABLE 8.2   Gasoline Demand Probabilities

Demand (Gallons)	Probabilities
1,000	0.01
1,150	0.02
1,250	0.05
1,350	0.12
1,450	0.2
1,550	0.27
1,650	0.18
1,750	0.08
1,850	0.04
2,000	0.03
Total Days =	1

Because cumulative probabilities will be more useful, we convert to a cumulative distribution function (CDF) as displayed in Table 8.3.

8.2.3 Inventory Algorithm

Inputs: Q = Delivery quantity in gallons

T = Time between deliveries in days

D = Delivery cost in dollars per delivery

S = Storage costs in dollars per gallons

N = Number of days in the simulation

TABLE 8.3 Gasoline CDF Demand

Demand (Gallons)	Probabilities
1,000	0.01
1,150	0.03
1,250	0.08
1,350	0.2
1,450	0.4
1,550	0.67
1,650	0.85
1,750	0.93
1,850	0.97
2,000	1

Output: C = Average daily cost

Step 1.   Initialize: Inventory $\rightarrow I = 0$ and $C = 0$.

Step 2.   Begin the next cycle with a delivery:

$$I = I + Q$$

$$C = C + D$$

Step 3.   Simulate each day of the cycle.

For $i = 1, 2, \ldots, T$, do steps 4–6.

Step 4.   Generate a demand, q_i. Use cubic splines to generate a demand based on a random CDF value, x_i.

Step 5.   Update the inventory: $I = I - q^i$.

Step 6.   Calculate the updated cost: $C = C + s * I$ if the inventory is positive.

If the inventory is ≤ 0, then set $I = 0$ and go to step 7.

Step 7.   Return to step 2 until the simulation cycle is completed.

Step 8.   Compute the average daily cost: $C = C/n$.

Step 9.   Output C. Stop.

We run the simulation and find that the average cost is about $5,753.04, and the inventory on hand is about 199,862.4518 gallons (or 199,863 gallons).

8.3 QUEUING MODEL

A queue is a waiting line, typically for some type of service. An example would be people in line to purchase a movie ticket or in a drive-through line to order fast food. There are two important entities in a queue: Customers and servers. There are some important parameters to describe a queue:

1. The number of servers available.

2. Customer arrival rate: Average number of customers arriving to be serviced in a time unit.

3. Server rate: Average number of customers processed in a time unit.

4. Time.

In many simple queuing simulations, as well as theoretical approaches, assume that arrivals and service times are exponentially distributed with a mean arrival rate of λ_1 and a mean service time of λ_2.

Theorem 8.1: If the arrival rate is exponential and the service rate is given by any distribution, then the expected number of customers waiting in line, L_q, and the expected waiting time, W_q, are given by

$$L_q = \frac{\lambda^2 \sigma^2 + \rho^2}{2(1-\rho)} \text{ and } W_q = \frac{L_q}{\lambda}.$$

where λ is the mean number of arrival per time period; μ is the mean number of customers serviced per time unit, $\rho = \lambda/\mu$, and σ is the standard deviation of the service time.

Here, we have a barber shop where we have two customers who arrive every 30 minutes. The service rate of the barber is three customers every 60 minutes. This implies the time between arrivals is 15 minutes, and the mean service time is one customer every 20 minutes. How many customers will be in the queue and what is their average waiting time?

8.3.1 Possible Solution with Simulation

We provide an algorithm for use.

8.3.1.1 Algorithm

For each customer 1 ... N,

Step 1. Generate an interarrival time, an arrival time, start time based on finish time of the previous customer, service time, completion time, amount of time waiting in a line, cumulative wait time, average wait time, number in queue, average queue length.

Step 2. Repeat N times.

Step 3. Compute output average wait time and queue length.

Stop

You will be asked to calculate the theoretical solution in the exercise set. We illustrate the simulation.

We will use the following to generate exponential random numbers, $x = -1/\lambda \ln(1 - \text{rand}())$

We generate a sample of 5,000 runs and plot customers versus average weight time (Figures 8.2–8.4).

D	E	F	G	H
Customer number	Time between arrivals	Arrival time	Start time	Service time
1	=-(1/$B$1)*LN(1-RAND())	=E2	=F2	=-(1/$B$2)*LN(1-RAND())
=D2+1	=-(1/$B$1)*LN(1-RAND())	=F2+E3	=MAX(I2,F3)	=-(1/$B$2)*LN(1-RAND())
=D3+1	=-(1/$B$1)*LN(1-RAND())	=F3+E4	=MAX(I3,F4)	=-(1/$B$2)*LN(1-RAND())
=D4+1	=-(1/$B$1)*LN(1-RAND())	=F4+E5	=MAX(I4,F5)	=-(1/$B$2)*LN(1-RAND())
=D5+1	=-(1/$B$1)*LN(1-RAND())	=F5+E6	=MAX(I5,F6)	=-(1/$B$2)*LN(1-RAND())
=D6+1	=-(1/$B$1)*LN(1-RAND())	=F6+E7	=MAX(I6,F7)	=-(1/$B$2)*LN(1-RAND())
=D7+1	=-(1/$B$1)*LN(1-RAND())	=F7+E8	=MAX(I7,F8)	=-(1/$B$2)*LN(1-RAND())
=D8+1	=-(1/$B$1)*LN(1-RAND())	=F8+E9	=MAX(I8,F9)	=-(1/$B$2)*LN(1-RAND())
=D9+1	=-(1/$B$1)*LN(1-RAND())	=F9+E10	=MAX(I9,F10)	=-(1/$B$2)*LN(1-RAND())

I	J	K	l
Completion time	wait time	Cumulative wait time	
=G2+H2	=G2-F2	=J2	=K2/D2
=G3+H3	=G3-F3	=K2+J3	=K3/D3
=G4+H4	=G4-F4	=K3+J4	=K4/D4
=G5+H5	=G5-F5	=K4+J5	=K5/D5
=G6+H6	=G6-F6	=K5+J6	=K6/D6
=G7+H7	=G7-F7	=K6+J7	=K7/D7

FIGURE 8.2   Excel Screenshot of Customers' Average Wait Time.

		Customer number	Time between arrivals	Arrival time	Start time	Service time	Completion time	Wait time	Cumulative wait time	Average wait
		1	1.934408754	1.934408754	1.9344088	0.071524668	2.005933422	0	0	0
		2	0.116601281	2.051010035	2.05101	0.714947959	2.765957994	0	0	0
		3	0.055768834	2.106778869	2.765958	0.36811946	3.134077454	0.659179	0.659179125	0.219726375
		4	0.879801355	2.986580224	3.1340775	0.206478939	3.340556393	0.147497	0.806676355	0.201669089
		5	0.095844504	3.082424728	3.3405564	1.055590069	4.396146462	0.258132	1.06480802	0.212961604
		6	1.043432803	4.125857531	4.3961465	0.63308224	5.029228702	0.270289	1.335096951	0.222516159
		7	0.223185659	4.34904319	5.0292287	0.818579146	5.847807847	0.680186	2.015282463	0.287897495
		8	2.251848324	6.600891514	6.6008915	0.393204228	6.994095741	0	2.015282463	0.251910308
		9	0.384299775	6.985191288	6.9940957	0.320344496	7.314440237	0.008904	2.024186916	0.224909657
		10	0.163595249	7.148786537	7.3144402	0.066657268	7.381097506	0.165654	2.189840616	0.218984062
		11	0.000502847	7.149289384	7.3810975	0.792646337	8.173743842	0.231808	2.421648738	0.220149885
		12	0.102456472	7.251745856	8.1737438	1.062486891	9.236230734	0.921998	3.343646724	0.278637227
		13	0.384817067	7.636562923	9.2362307	0.2167925	9.453023234	1.599668	4.943314534	0.380254964
		14	0.625581112	8.262144036	9.4530232	0.342525761	9.795548994	1.190879	6.134193732	0.438156695
		15	0.5489886	8.811132636	9.795549	0.392117518	10.18766651	0.984416	7.118610091	0.474574006
		16	0.540099845	9.351232481	10.187667	0.500628779	10.68829529	0.836434	7.955044122	0.497190258
		17	0.025796165	9.377028647	10.688295	0.051349505	10.7396448	1.311267	9.266310766	0.545077104
		18	0.199860228	9.576888875	10.739645	0.497924558	11.23756935	1.162756	10.42906669	0.579392594
		19	0.422003799	9.998892674	11.237569	0.610221593	11.84779095	1.238677	11.66774337	0.614091756
		20	1.086641979	11.08553465	11.847791	0.139853034	11.98764398	0.762256	12.42999966	0.621499983
		21	0.085067941	11.17060259	11.987644	0.304856673	12.29250065	0.817041	13.24704105	0.630811478
		22	0.688452558	11.85905515	12.292501	0.13728232	12.42978297	0.433446	13.68048655	0.621840298

FIGURE 8.3 Customers' Average Wait Time.

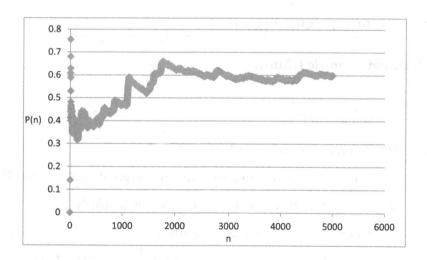

FIGURE 8.4 Graph of Average Wait Times.

We note that the plot appears to be converging at values slightly higher than 0.6666. Thus, we will run 100 more trials of 5,000 and compute the average.

We obtain the descriptive statistics from Excel. We note the mean is 0.6601 that is very close to our theoretical mean.

TABLE 8.4   Descriptive Statistics from Excel

Descriptive Statistics	
Mean	0.660
Standard error	0.006
Median	0.658
Mode	#N/A
Standard deviation	0.063
Sample variance	0.004
Kurtosis	−0.319
Skewness	0.156
Range	0.319
Minimum	0.501
Maximum	0.819
Sum	66.015
Count	100.000

8.4 R APPLIED SIMULATION

Algorithm: Example 1. Missiles

INPUTS:

N = Number of F-15s

M = Number of missiles fired, M = 10

P = Probability the one F-15 can destroy the target, P = 0.9 × 0.5 = 0.45

Q = Probability that one air-defense missile can disable the F-15 Q = 0.6

X = Number of F-15 disabled in the attack

Pi = Probability of mission success given X = $i \rightarrow 1 - (1 - P)^{N-i}$

B = Binomial probability, given X = $i \binom{m}{i} q^i (1-q)^{m-i}, i = 0, 1, 2, \ldots, m$

$S = \sum P_i B_i$

OUTPUT   S = Probability of mission success

Step 1. Initialize S = 0

Step 2. For i = 0 to M do

Step 3. $P(i) = 1 - (1 - P)^{N - I}$

Step 4. B(i) = Binomial Distribution for (m, *i*, q)

Step 5. Compute S = S + P(*i*) × B(*i*)

Step 6. Output S

Step 7. Stop

R Code

```
S=0
> n=c(10,11,12,13,14,15,16,17,18,19,20)
> for (i in 0:10) {
+ pn= 1-(1-.45)^(n-i)
+ x=rbinom(i,10,.45)
+ xs=dbinom(i,10,.6)
+ s=s+pn*xs
+ print (s)}
[1] 0.0001045920 0.0001047115 0.0001047773 0.0001048134 0.0001048333
[6] 0.0001048442 0.0001048502 0.0001048536 0.0001048554 0.0001048564
[11] 0.0001048569
[1] 0.001670212 0.001673592 0.001675450 0.001676472 0.001677034
[6] 0.001677344 0.001677514 0.001677607 0.001677659 0.001677687
[11] 0.001677703
[1] 0.01219815 0.01224153 0.01226539 0.01227851 0.01228573 0.01228970
[7] 0.01229188 0.01229309 0.01229375 0.01229411 0.01229431
[1] 0.05401894 0.05435326 0.05453714 0.05463827 0.05469390 0.05472449
[7] 0.05474132 0.05475057 0.05475566 0.05475846 0.05476000
[1] 0.1624099 0.1641328 0.1650804 0.1656016 0.1658883 0.1660459
[7] 0.1661326 0.1661803 0.1662066 0.1662210 0.1662289
[1] 0.3529692 0.3592366 0.3626837 0.3645796 0.3656223 0.3661958
[7] 0.3665112 0.3666847 0.3667801 0.3668326 0.3668615
[1] 0.5808401 0.5974358 0.6065634 0.6115836 0.6143447 0.6158633
[7] 0.6166986 0.6171579 0.6174106 0.6175496 0.6176260
[1] 0.7600618 0.7927536 0.8107341 0.8206234 0.8260625 0.8290540
[7] 0.8306993 0.8316042 0.8321019 0.8323757 0.8325262
[1] 0.8444121 0.8935658 0.9206004 0.9354694 0.9436473 0.9481452
[7] 0.9506190 0.9519796 0.9527280 0.9531396 0.9533659
[1] 0.8625520 0.9216826 0.9542045 0.9720915 0.9819293 0.9873402
[7] 0.9903161 0.9919529 0.9928531 0.9933482 0.9936205
[1] 0.8625520 0.9244036 0.9584220 0.9771321 0.9874226 0.9930825
[7] 0.9961954 0.9979074 0.9988491 0.9993670 0.9996519
> print (s)
[1] 0.8625520 0.9244036 0.9584220 0.9771321 0.9874226 0.9930825
[7] 0.9961954 0.9979074 0.9988491 0.9993670 0.9996519
>
N= 15 is the first time our probability of success is greater than 99% at 99.30825%.
```

We find that the number of F-15s equaling 15 gives us P(s) = 0.9930825.

Any number of F-15 greater than 15 works to provide a result with the probability of success we desire. We would think the 15 F-15s yielding a P(s) = 0.9930825 would suffice. Any more would be overkill.

8.5 EXERCISES

8.1. Tollbooths: Heavily traveled toll roads such as the Garden State Parkway, Interstate 95, and so forth are multilane divided highways that are interrupted at intervals by toll plazas. Because collecting tolls is usually unpopular, it is desirable to minimize motorist annoyance by limiting the amount of traffic disruption caused by the toll plazas. Commonly, a much larger number of tollbooths are provided than the number of travel lanes entering the toll plaza. On entering the toll plaza, the flow of vehicles fans out to the larger number of tollbooths; when leaving the toll plaza, the flow of vehicles is forced to squeeze down to a number of travel lanes equal to the number of travel lanes before the toll plaza. Consequently, when traffic is heavy, congestion increases when vehicles leave the toll plaza. When traffic is very heavy, congestion also builds at the entry to the toll plaza because of the time required for each vehicle to pay the toll.

Construct a mathematical model to help you determine the optimal number of tollbooths to deploy in a barrier-toll plaza. Explicitly, first consider the scenario in which there is exactly one tollbooth per incoming travel lane. Then consider multiple tollbooths per incoming lane. Under what conditions is one tollbooth per lane more or less effective than the current practice? Note that the definition of optimal is up to you to determine.

8.2. Major League Baseball: Build a simulation to model a baseball game. Use your two favorite teams or favorite all-star players to play a regulation game.

8.3. Automobile Emissions: Consider a large engineering company that performs emissions control inspections on automobiles for the state. During the peak period, cars arrive at a single location that has four lanes for inspections following exponential arrivals with a mean of 15 minutes. Service times during the same period are uniform: Between [15,30] minutes. Build a simulation for the length of the queue. If cars wait more than 1 hour, the company pays a penalty of $200 per car. How much money, if any, does the company pay in penalties? Would more inspection lanes help? What costs associated with the inspection lanes need to be considered?

8.6 RECRUITING SIMULATION MODEL

Monthly demand for recruits is provided in the following table.

Demand	Probability	CDF
400	0.05	0.05
420	0.10	0.15
440	0.25	0.40
460	0.30	0.70
480	0.25	0.95
500	0.05	1.0

Additionally, depending on conditions, the average cost per recruit is between \$70 and \$90 in integer values. Returns from Higher HQ are between 20% and 30% of costs. There is a fixed cost of \$2,500/month for the office, phones, etc. Build a simulation model to determine the average monthly costs.

Assume Cost = Demand × Cost per Recruit + Fixed Cost-Return Amount,

where Return = % × Cost

8.7 SIMPLE QUEUING PROBLEM

The store manager is trying to improve customer satisfaction by offering better service. They want the average customer to wait less than 3 minutes and the average length of the queue (line) is 2 or fewer. The store estimates about 180 customers per day. The existing service and arrival times are given.

Service Time	Probability	Time between Arrival	Probability
1	0.30	0	0.10
2	0.20	1	0.15
3	0.35	2	0.15
4	0.15	3	0.35
		4	0.25
		5	0.05

Determine if the current servers are satisfying the goals. If not, how much improvement is needed in service to accomplish the stated goals.

REFERENCES AND ADDITIONAL READINGS

Giordano, F., Fox, W., and Horton, S. *A First Course in Mathematical Modeling.* 5th Ed. Boston, MA: Cengage Publishers. 2014.

Law, A., and Kelton, D. *Simulation Modeling and Analysis.* 4th Ed. New York: McGraw Hill. 2007.

Meerschaert, M. M. *Mathematical Modeling.* San Diego: Academic Press. 1993.

Winston, W. *Operations Research: Applications and Algorithms.* 3rd Ed. Belmont, CA: Duxbury Press. 1994.

CHAPTER 9

System Reliability Modeling

RELIABILITY MODELS ARE MATHEMATICAL tools that help decision-makers estimate the probability of failure, the mean time to failure, and the availability of the system. They allow practitioners to describe and understand the ability of the system or individual components of the system to function under stated conditions for some time. Reliability models can help decision-makers optimize resource allocations, reduce risk and costs, or develop maintenance strategies. This chapter is adapted from Military Reliability Modeling by Fox and Horton (1992).

9.1 INTRODUCTION TO RELIABILITY MODELING

You are a New York City policeperson. You are on a stakeout and must occupy a position for at least the next 24 hours. Hourly situation reports must be made by radio. All necessary food, equipment, and supplies for the 24-hour period must be carried with the police. The stakeout is ineffective unless it can communicate with you in a timely manner. Therefore, radio communications must be reliable. The radio has several components which affect its reliability, an essential one being the battery.

Batteries have a useful life which is not deterministic (we do not know exactly how long a battery will last when we install it). Its lifetime is a variable which may depend on previous use, manufacturing defects, weather, etc. The battery that is installed in the radio prior to leaving for the stakeout could last only a few minutes or for the entire 24 hours. Since communications are so important to this mission, we are interested in modeling and analyzing the reliability of the battery.

DOI: 10.1201/9781003464969-9

We will use the following definitions for reliability.

9.1.1 Definition

If T is the time to failure of a component of a system, and f (t) is the probability distribution function of T, then the components' reliability at time t is R $(t) = P(T > t) = 1 - F(t)$. R (t) is called the reliability function, and F (t) is the cumulative distribution function of f (t).

A measure of this reliability is the probability that a given battery will last more than 24 hours. If we know the probability distribution for the battery life, we can use our knowledge of probability theory to determine the reliability. If the battery reliability is below acceptable standards, one solution is to have the police persons carry spares. Clearly, the more spares they carry, the less likely there is to be a failure in communications due to batteries. Of course, the battery is only one component of the radio. Others include the antenna, handset, etc. Failure of any one of the essential components causes the system to fail. This is a relatively simple example of one of many military applications of reliability.

This chapter will demonstrate how you can use elementary probability to generate models that can be used to determine the reliability of equipment and systems.

9.2 MODELING COMPONENT RELIABILITY

In this section, we will discuss how to model component reliability. The reliability function, $R(t)$, is defined as:

$$R(t) = P(T > t) = P \text{ (component fails after time } t).$$

This can also be stated, using T as the component failure time, as

$$R(t) = P(T > t) = - P(T \le t) = 1 - \int_{-\infty}^{t} f(x)f(x)dx = 1 - F(t)$$

Thus, if we know the probability density function $f(t)$ of the time to failure T, we can use probability theory to determine the reliability function $R(t)$. We normally think of these functions as being time dependent; however, this is not always the case. The function might be discrete such as the lifetime of a cannon tube. It is dependent on the number of rounds fired through the tube (a discrete random variable).

A useful probability distribution in reliability is the exponential distribution. Recall that its density function is given by

$$f(t) = \begin{cases} \lambda e^{-\lambda t} & t \geq 0 \\ 0 & otherwise \end{cases}$$

where the parameter is λ.

We know λ is such that it's reciprocal, $\frac{1}{\lambda}$, equals the mean of the random variable, T. If T denotes the time to failure of a piece of equipment or a system, then $\frac{1}{\lambda}$ is the mean time to failure which is expressed in units of time. For applications of reliability, we will use the parameter λ. Since $\frac{1}{\lambda}$ is the mean time to failure, λ is the average number of failures per unit time or the failure rate. For example, if a light bulb has a time to failure that follows an exponential distribution with a mean time to failure of 50 hours, then its failure rate is 1 light bulb per 50 hours or 1/50 per hour, so in this case $\lambda = 0.02$ per hour. Note that the mean of T, the mean time to failure of the component, is 1/50.

9.2.1 Example 9.1: Battery Problem – Reliability

Let's consider the example presented in the introduction. Let the random variable T be defined as follows: T = Time until a randomly selected battery fails.

Suppose radio batteries have a time to failure that are exponentially distributed with a mean of 30 hours. In this case, we could write:

$$T \sim \exp(\lambda = \frac{1}{30})$$

Therefore, $\lambda = 1/30$ so that we know the density function is

$F(t) = 1/30\ e^{-t/30}$ for $t \geq 0$ and F (t) is as follows:

$$F(t) = \int_0^t \frac{1}{30} e^{-x/30} dx$$

$F(t)$, the CDF of the exponential distribution, can be integrated to obtain

$$1 - e^{(-t/30)} \text{ for } t \geq 0$$

$$F(t) = 1 - e^{-t/30}, t > 0$$

Now we can compute the reliability function for a battery: Recall that in the earlier example, the police must occupy the stakeout for 24 hours. The reliability of the battery for 24 hours is

$$R(24) = 1 - (1 - e^{(-24/30)}) = 1 - 0.55067 = 0.44933$$

So, the probability that the battery lasts more than 24 hours is 0.4493.

9.2.2 Example 9.2: Battery Problem Revisited – Reliability

We have the option to purchase a new nickel cadmium battery for our stakeout. Testing has shown that the distribution of the time to failure can be modeled using a parabolic function:

$$f(x) = \begin{cases} \dfrac{x}{384}\left(1 - \dfrac{x}{48}\right) & 0 \leq x \leq 48 \\ 0 & \textit{otherwise} \end{cases}$$

Let the random variable T be defined as follows:

T = Time until a randomly selected battery fails.

In this case, we could write

$$f(t) = (t/384)(1 - t/48),\ 0 \leq t \leq 48,$$

$$\text{and } F(t) = \int_0^{} (x/384)(1 - x/48)dx.$$

Recall that in the earlier example the persons must be on the stakeout for 24 hours. The reliability of the battery for 24 hours is therefore:

$$R(24) = 1 - F(24) = 1 - \int (t/384)(1 - t/48)dx = 0.500,$$

which is an improvement over the batteries from Example 9.1. Therefore, we should use the new battery.

9.3 MODELING SERIES AND PARALLEL COMPONENTS

9.3.1 Modeling Series Systems

Now we consider a system with n components $C_1, C_2 \ldots, Cn,$ where each of the individual components must work for the system to function. A model of this type of system is shown in Figure 9.1.

If we assume these components are mutually independent, the reliability of this type of system is easy to compute. We denote the reliability of component i at time t by Ri (t). In other words, Ri (t) is simply the probability that component i will function continuously from time 0 through until time t. We are interested in the reliability of the entire system of n components, but since these components are mutually independent, the system reliability is:

$$R(t) = R_1(t) \cdot R_2 (t) \cdots \cdot R_n (t).$$

9.3.2 Example 9.3: Radio Components

Our radio has several components. Let us assume that there are four major components – they are (in order) the handset, the battery, the receiver-transmitter, and the antenna. Since they all must function properly for the radio to operate, we can model the radio with the diagram shown in Figure 9.2.

Suppose we know that the probability that the handset will work for at least 24 hours is 0.6703, and the reliabilities for the other components are 0.4493, 0.7261, and 0.9531, respectively. If we assume that the components work *independently* of each other, then the probability that the entire system works for 24 hours is:

$$R(24) = R_1(24) \cdot R_2 (24) \cdot R_3 (24) \cdot R_4 (24) = (.6703)(.4493)$$
$$(.7261)(.9531) = 0.2084.$$

Two events A and B are *independent* if $P(A|B) = P(A)$.

FIGURE 9.1   Series System.

9.3.3 Modeling Parallel Systems (Two Components)

Now we consider a system with two components where only one of the components must work for the system to function. A system of this type is depicted in Figure 9.3.

Notice that in this situation, the two components are *both* put in operation at time 0; they are both subject to failure throughout the period of interest. Only when *both* components fail before time t does the system fail. Again we also assume that the components are independent. The reliability of this type of system can be found using the following well addition known model:

$$P(A \cup B) = P(A) + P(B) - P(A \cap B).$$

In this case, A is the event that the first component functions for longer than some time, t, and B is the event that the second component functions longer than the same time, t. Since reliabilities *are* probabilities, we can translate the above formula into the following:

$$R(t) = R_1(t) + R_2(t) - R_1(t) R_2(t).$$

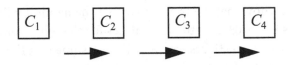

FIGURE 9.2   Radio System.

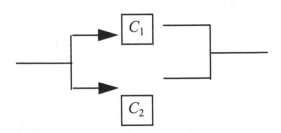

FIGURE 9.3   Parallel System of Two Components.

9.3.4 Example 9.4: Parallel Bridges

Suppose we have two bridges in the area for a Boy Scout hike. It will take 3 hours to complete the crossing for all the hikers. The crossing will be successful as long as at least one bridge remains operational during the entire crossing period. You estimate that the bridges are in bad shape and that bridge 1 has a one-third chance of being destroyed and a one-fourth chance of destroying bridge 2 in the next 3 hours. Assume the destruction of the bridges is independently done. What is the probability that your boy scouts can complete the crossing?

Solution: First, we compute the individual reliabilities:

$$R_1(3) = 1 - 1/3 = 2/3$$

and

$$R_2(3) = 1 - 1/4 = 3/4$$

Now it is easy to compute the system reliability for 3 hours:

$$R(System) = R_1(3) + R_2(3) - R_1(3)R_2(3) = 2/3 + \frac{3}{4} - (2/3)(3/4) = 0.9166667$$

9.4 MODELING ACTIVE REDUNDANT SYSTEMS

Consider the situation in which a system has n components, all of which begin operating (are active) at time $t = 0$. The system continues to function properly as long as at least k of the components do not fail. In other words, if $n \le k \le 1$ components fail, the system fails. This type of component system is called an active redundant system. The active redundant system can be modeled as a parallel system of components as shown in Figure 9.4:

We assume that all n components are identical and will fail independently.

If we let T_i be the time to failure of the ith component, then the T_i terms are independent and identically distributed for $i = 1, 2, 3, \ldots, n$. Thus, $Ri(t)$, the reliability at time t for component i, is identical for all components.

Recall here that our system operates if at least k components function properly.

Now we define the random variables X and T as follows:

X = Number of components functioning at time t.

T = Time to failure of the entire system.

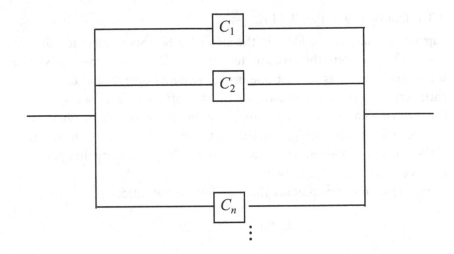

FIGURE 9.4   Active Redundant System.

Then, we have

$$R(t) = P(T > t) = P(X \geq k).$$

It is easy to see that we now have n identical and independent components with the same probability of failure by time t. This situation corresponds to a binomial experiment, and we can solve for the system reliability using the binomial distribution with parameters n and $p = Ri\ (t)$.

9.4.1 Example 9.5: Manufacturing

A manufacturing company has 15 different machines to make item A. They estimate if at least 12 are operating, they will be able to make all the items necessary to meet demand. Machines are assumed to be in parallel (active redundant), i.e.: They fail independently. If we know that each machine has a 0.6065 probability of operating properly for at least 24 hours, we can compute the reliability of the entire machine system for 24 hours.

Define the random variable: X = Number of machines working after 24 hours.

Clearly, the random variable X is binomially distributed with $n = 15$ and $p = 0.6065$. In the language of mathematics, we write this sentence as:

$$X \sim b(15, 0.6065) \text{ or } X \sim \text{BINOMIAL}(15, 0.6065).$$

We know that the reliability of the machine system for 24 hours is:

$$R(24) = P(X \geq 12) = P(12 \leq X \leq 15) = 0.0990.$$

Thus, the reliability of the system for 24 hours is only 0.0990.

9.5 MODELING STANDBY REDUNDANT SYSTEMS

Active redundant systems can sometimes be inefficient. These systems require only k of the n components to be operational, but all n components are initially in operation and thus subject to failure. An alternative is the use of spare components. Such systems have only k components initially in operation; exactly what we need for the whole system to be operational. When a component fails, we have a spare "standing by" which is immediately put in to operation. For this reason, we call these *Standby Redundant Systems*. Suppose our system requires k operational components and we initially have $n \leq k$ spares available. When a component in operation fails, a decision switch causes a spare or standby component to activate (becoming an operational component). The system will continue to function until there are less than k operational components remaining. In other words, the system works until $n - k + 1$ components have failed. We will consider only the case where one operational component is required (the special case where $k = 1$) and there are $n - 1$ standby (spare) components available. We will assume that a decision switch (DS) controls the activation of the standby components instantaneously and 100% reliably. We use the model shown in Figure 9.5 to represent this situation.

If we let T_i be the time to failure of the ith component, then the T_i's are independent and identically distributed for $i = 1, 2, 3, \ldots, n$. Thus, $Ri\ (t)$ is identical for all components. Let $T =$ Time to failure of the entire system. Since the system fails only when all n components have failed, and component $i + 1$ is put into operation only when component i fails, it is easy to see that:

$$T = T_1 + T_2 + \ldots + T_n.$$

In other words, we can compute the system failure time easily if we know the failure times of the individual components.

Finally, we can define a random variable $X =$ Number of components that *fail* before time t in a standby redundant system. Now the reliability of

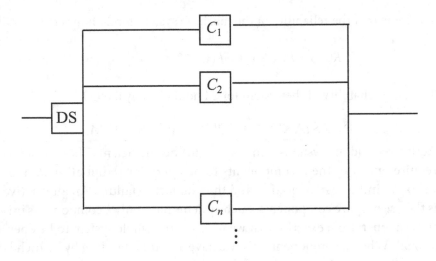

FIGURE 9.5   Standby Redundant System.

the system is simply equal to the probability that less than n components fail during the time interval $(0, t)$. In other words,

$$R(t) = P(X < n).$$

It can be shown that X follows a Poisson distribution with parameter $\lambda = \alpha t$ where α is the failure rate, so we write:

$$X \sim \text{POISSON}(\lambda).$$

For example, if time is measured in seconds, then α is the number of failures per second. The reliability for some specific time t then becomes:

$$R(t) = P(X < n) = P(0 \le X \le n - 1).$$

9.5.1 Example 9.6: Battery Problems Revisited for Standby

Consider the reliability of a radio battery. We determined previously that one battery has a reliability for 24 hours of 0.4493. Considering the importance of communications, you decide that this reliability is not satisfactory. Suppose we carry two spare batteries. The addition of the spares should increase the battery system reliability.

Later in the course, you will learn how to calculate the failure rate α for a battery, given the reliability (0.4493 in this case). For now, we will give this to you: $\alpha = 1/30$ per hour. We know that $n = 3$ total batteries.

Therefore:

$$X \sim Poisson(\lambda = \alpha t = 24/30 = 0.8) \text{ and}$$

$$R(24) = P(X{<}3) = P(0 \le X \le 2) = 0.9526$$

The reliability of the system with two spare batteries for 24 hours is now 0.9526.

9.5.2 Example 9.7: Stakeout Problem Revisited

If the police stakeout must stay out for 48 hours without resupply, how many spare batteries must be taken to maintain a reliability of 0.95? We can use trial and error to solve this problem. We start by trying our current load of 2 spares. We have:

$$X \sim Poisson(\lambda = \alpha t = \frac{48}{30} = 1.6)$$

and we can now compute the system reliability

$$R(48) = P(X < 3) = P(0 \le X \le 2) = 0.7834 < 0.95$$

which is not good enough. Therefore, we try another spare so $n < 4$ (3 spares) and we compute:

$$R(48) = P(X < 4) = P(0 \le X \le 3) = 0.9212 < 0.95$$

which is still not quite good enough, but we are getting close! Finally, we try $n < 5$ which turns out to be sufficient:

$$R(48) = P(X < 5) = P(0 \le X \le 4) = 0.9763 \ge 0.95.$$

Therefore, we conclude that the stakeout should take out at least 4 spare batteries for a 48-hour mission.

9.6 MODELS OF LARGE-SCALE SYSTEMS

In our discussion of reliability up to this point, we have discussed series systems, active redundant systems, and standby redundant systems. Unfortunately, things are not always this simple. The types of systems listed above often appear as subsystems in larger arrangements of components that we shall call "large-scale systems". Fortunately, if you know how to deal with series systems, active redundant systems, and standby redundant systems, finding system reliabilities for large-scale systems is easy.

The first and most important step in developing a model to analyze a large-scale system is to draw a picture. Consider the network that appears as Figure 9.6. Subsystem A is the standby redundant system of three components (each with failure rate 5 per year) with the decision switch on the left of the figure. Subsystem B_1 is the active redundant system of three components (each with failure rate 3 per year), where at least two of the three components must be working for the subsystem to work. Subsystem B_1 appears in the upper right portion of the figure. Subsystem B_2 is the two-component parallel system in the lower right portion of the figure. We define subsystem B as being subsystems B_1 and B_2 together. We assume all components have exponentially distributed times to failure with failure rates as shown in Figure 9.6.

Suppose we want to know the reliability of the whole system for 6 months. Observe that you already know how to compute reliabilities for the subsystems A, B_1, and B_2. Let's review these computations and then see how we can use them to simplify our problem.

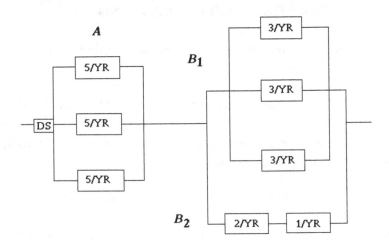

FIGURE 9.6   Network Example.

Subsystem A is a standby redundant system, so we will use the Poisson model. We let

X = The number of components which fail in one year.

Since 6 months is 0.5 years, we seek $R_A (0.5) = P(X < 3)$ where X follows a Poisson distribution with parameter $11 = \alpha t = (5)(0.5) = 2.5$. Then,

$$R_A (0.5) = P(X < 3) = P(0 \leq X \leq 2) = 0.5438.$$

Now we consider subsystem B_1. In Section 9.2, we learned how to find individual component reliabilities when the time to failure followed an exponential distribution. For subsystem B_1, the failure rate is 3 per year, so our individual component reliability is

$$R(0.5) = 1 - F(0.5) = 1 - (1 - e^{-(3)(0.5)}) = e^{-(3)(0.5)} = 0.2231.$$

Now recall that subsystem B_1 is an active redundant system where two components of the three must work for the subsystem to work. If we let

Y = The number of components that function for 6 months

and recognize that Y follows a binomial distribution with $n = 3$ and $p = 0.2231$, we can quickly compute the reliability of the subsystem B_1 as follows:

$$R_B (0.5) = P(Y \geq 2) = 1 - P(Y < 2) = 1 - P(Y \leq 1) = 1 - 0.8729 = 0.1271.$$

1

Finally, we can look at subsystem B_2. Again, we use the fact that the failure times follow an exponential distribution. The subsystem consists of two components; obviously they both need to work for the subsystem to work. The first component's reliability is

$$R(0.5) = 1 - F(0.5) = 1 - (1 - e^{-(2)(0.5)}) = e^{-(2)(0.5)}$$

$$= 0.3679,$$

and for the other component, the reliability is

$$R(0.5) = 1 - F(0.5) = 1 - {}_(1 - \varepsilon^{-(1)(0.5)}) = e^{-(1)(0.5)}$$

$$= 0.6065.$$

Therefore, the reliability of the subsystem is:

$$R_B(0.5) = (0.3679)(0.6065) = 0.2231.$$

Our overall system can now be drawn as shown in Figure 9.7.

From here, we determine the reliability of subsystem B by treating it as a system of two independent components in parallel where only one component must work. Therefore,

$$R_B(0.5) = R_{B1}(0.5) + R_{B2}(0.5) - R_{B1}(0.5) \cdot R_{B2}(0.5)$$

$$= 0.1271 + 0.2231 - (0.1271)(0.2231) = 0.3218$$

Finally, since subsystems A and B are in series, we can find the overall system reliability for 6 months by taking the product of the two subsystem reliabilities:

$$R_{system}(0.5) = R_A(0.5) \cdot R_B(0.5) = (0.5438)(0.3218) = 0.1750.$$

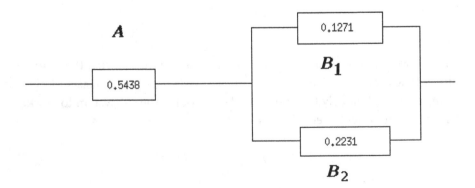

FIGURE 9.7 Simplified Network Example.

We have used a network reduction approach to determine the reliability for a large-scale system for a given time period. Starting with those subsystems which consist of components independent of other subsystems, we reduced the size of our network by evaluating each subsystem reliability one at a time. This approach works for any large-scale network consisting of basic subsystems of the type we have studied (series, active redundant, and standby redundant).

We have seen how methods from elementary probability can be used to model military reliability problems. The modeling approach presented here is useful in helping students simultaneously improve their understanding of both the military problems addressed and the mathematics behind these problems. The models presented also motivate students to appreciate the power of mathematics and its relevance to today's world.

9.7 EXERCISES

9.1. A continuous random variable Y, representing the time to failure of a 0.50 mm tube, has a probability density function given by

$$f(y) = \begin{cases} 1/3e^{-y/3} & y \geq 0 \\ 0 & otherwise \end{cases}$$

  a. Find the reliability function for Y.

  b. Find the reliability for 1.2 time periods, $R(1.2)$.

9.2. The lifetime of a car engine (measured in time of operation) is exponentially distributed with a MTTF of 400 hours. You have received a mission that requires 12 hours of continuous operation. Your log book indicates that the car engine has been operating for 158 hours.

  a. Find the reliability of your engine for this mission.

  b. If your vehicle's engine had operated for 250 hours prior to the mission, find the reliability for the mission.

9.3. A criminal must be captured. You are on the police SWAT team when they decide to use helicopter to help capture the key criminal. The SWAT aviation battalion is tasked to send four helicopters. On their way to the target area, these helicopters must fly over foggy territory for approximately

15 minutes during which they have the risk of vulnerable accidents. The lifetime of a helicopter over this territory is estimated to be exponentially distributed with a mean of 18.8 minutes. It is further estimated that two or more helicopters are required to capture the criminal. Find the reliability of the helicopters in accomplishing their mission (assuming the only reason a helicopter fails to reach the target is the foggy weather).

9.4. For the mission in Exercise 9.3, the police commissioner determines that, to justify risking the loss of helicopter, there must be at least an 80% chance of capturing the criminal. How many helicopters the aviation battalion recommend be sent? Justify your answer.

9.5. Mines are a dangerous obstacle. Most mines have three components – the firing device, the wire, and the mine itself (casing). If any of these components fail, the systems fail. These components of the mine start to "age" when they are unpacked from their sealed containers. All three components have MTTF that are exponentially distributed of 60 days, 300 days, and 35 days, respectively.

a. Find the reliability of the mine after 90 days.

b. What is the MTTF of the mine?

c. What assumptions, if any, did you make?

9.6. You are a project manager for the new system being developed in Huntsville, Alabama. A critical subsystem has two components arranged in a parallel configuration. You have told the contractor that you require this subsystem to be at least 0.995 reliable. One of the subsystems came from an older system and has a known reliability of 0.95. What is the minimum reliability of the other component so that we meet our specifications?

9.7. You are in charge of stage lighting for an outdoor concert. There is some concern about the reliability of the lighting system for the stage. The lights are powered by a 1.5 KW generator that has a MTTF of 7.5 hours.

a. Find the reliability of the generator for 10 hours if the generator's reliability is exponential.

b. Find the reliability of the power system if two other identical 1.5 KW generators are available. First consider as active redundant and then as standby redundant. Which would improve the reliability the most?

c. How many generators would be necessary to ensure a 99% reliability?

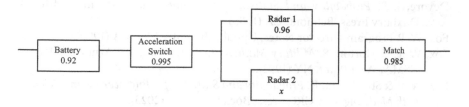

9.8. Consider the system below with the reliability for each component as indicated. Assume all components are independent and the radars are active redundant.

a. Find the system reliability for six months when $x = 0.96$.

b. Find the system reliability for six months when $x = 0.939$.

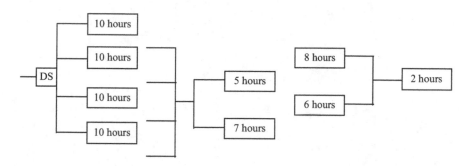

9.9. Major system has components as shown in diagram below. All components have exponential times to failure with mean times to failure shown. All components operate independently of each other. Find the reliability for this weapon system for 2 hours.

9.10. A system comprises three components. These components have constant failure rates of 14, 15, and 13 hours. The system will stop working, if any one of its components fails. Find the reliability of the system for 25 hours.

9.11. For a 2-out-of-3 system (a special case of k-out-of-n system), each component has reliability of 0.9. What is the system reliability?

9.12. A space shuttle requires 3-out-of-4 of its main engines to achieve orbit. Each engine has a reliability of 0.97. What is the probability of achieving orbit?

REFERENCES AND SUGGESTED READINGS

Devore, J. L. *Probability and Statistics for Engineers and Scientists*. 4th Ed. Duxbury Press, Belmont, CA (1995).

Fox, W. P. Program Director Notes, Unpublished MA 206, MA (1990).

Fox, W. P. & Horton, S. *Military Mathematical Modeling*. D. Arney ed. 1st Ed. USMA, West Point, NY (1992).

Fox, W. P. & Sturdivant, R. *Probability and Statistics for Engineering and Sciences with Modeling in R*. CRC Press, Boca Raton, FL (2023).

Resnick, S. L. *Adventures in Stochastic Processes*. Birkhäuser, Boston (1992).

CHAPTER 10

Modeling Decision-Making with Multi-Attribute Decision Modeling with Technology

10.1 INTRODUCTION

This chapter updated from Fox and Burks (2021) describes the elements that make up a multi-attribute decision-making (MADM) problem and provides an overview of alternative MADM methods. Section 10.2 discusses the weighting methods. Sections 10.3–10.5 provide details and examples on how to use selected MADM methods, such as sum of additive weights (SAW), technique of order preference by similarity to ideal solution (TOPSIS), and data envelopment analysis (DEA).

MADM methods are techniques used to address problems where a decision-maker is trying to choose between alternatives or ranking a finite *number of alternatives* which are measured by two *or more relevant attributes*. There is a vast literature describing MADM theory, methods, and applications. We have provided a good source for additional information in the Additional Readings section.

DOI: 10.1201/9781003464969-10

10.1.1 Elements of a MADM Problem

All MADM problems have four common characteristics.

10.1.1.1 Finite (Generally Small) Set of Alternatives

The first is that in general all MADM problems involve a finite and small set of discrete and predetermined options or alternatives. In this way, MADM problems are distinguished from other decision processes, which involve the *design* of a "best" alternative by considering the trade-offs within a set of interacting design constraints.

10.1.1.2 Trade-Offs among Attributes

Second, MADM problems occur when there is no single alternative or the most preferred available value or performance for all attributes. This is often the result of an underlying trade-off relationship among attributes. One example may be the trade-off between low desired energy costs and large glass window areas (which may raise heating and cooling costs while lowering lighting costs and enhancing aesthetics).

10.1.1.3 Incommensurable Units

The attributes will generally not all be measurable in the same units in a MADM problem. In fact, some attributes may be either impractical, impossible, or too costly to measure at all. For example, life-cycle costs are directly measured in dollars, the number and size of offices are measured in other units, and the public image of a building may not be practically measurable in any unit. If all relevant attributes characterizing alternative buildings can be expressed in terms of financial costs or benefits scheduled to occur at specifiable times, then the ranking and selection of a building do not require the application of MADM. Often, we normalize these values so that they are both scaled and dimensionless.

10.1.1.4 Decision Matrix

A MADM problem can be characterized by a "decision matrix". We will describe this in more detail in Sections 10.2 and 10.3. The decision matrix indicates both the set of alternatives and the set of attributes being considered in each problem. It summarizes the "raw" data available to the decision-maker at the start of the analysis. A decision matrix has a row corresponding to each alternative being considered and a column corresponding to each attribute being considered. A problem with a total of m alternatives characterized by n attributes is described by an $m \times n$ matrix X.

MADM approaches are alternative methods for combining the information in a problem's decision matrix together with additional information from the decision-maker to determine a final ranking, screening, or selection from among the alternatives. Besides the information contained in the decision matrix, all but the simplest MADA techniques require additional information from the decision-maker in order to arrive at a final ranking, screening, or selection.

For example, the decision matrix provides no information about the relative importance of the different attributes to the decision-maker, nor about any minimum acceptable, maximum acceptable, or target values for particular attributes.

10.2 DELPHI METHOD

The Delphi Method is a reliable way of obtaining the opinions of a group of experts on an issue by conducting several rounds of interrogative communications. This method was first developed in the US Air Force in the 1950s, mainly for market research and sales forecasting (Fox 2018). This method is essentially a communication device that is particularly useful for achieving a consensus among experts, given a complex problem. The method consists of repeated solicitations of questions from a panel of experts who are anonymous. The information and ideas of each panel member are distributed among all the panel members in the next round. They can comment on others' viewpoints and can even use new information to modify their own opinions. Panel members can change their opinions based on new information more easily than in regular group meetings and open discussion. A consensus of opinions should be ultimately achieved in this way. This method will also highlight the areas where panel members have disagreements or uncertainty in a quantitative manner. The evaluation of belief statements by the panel as a group is an explicit part of the Delphi Method (Fox 2018). The panel consists of a number of experts chosen based on their experience and knowledge. As mentioned previously, panel members remain anonymous to each other throughout the procedure in order to avoid the negative impacts of criticism on the innovation and creativity of panel members. The Delphi Method should be conducted by a director (facilitator) who has independent communication with each panel member. The director develops a questionnaire based on the problem at hand and sends it to each panel member. Then, the responses to the questions and all the comments are collected and evaluated by the director. The director should process the information and filter out the irrelevant content. The result of the processed information will again be distributed among the panel members. Each member receives new information and ideas and can comment on them and/or revise his own previous opinions. One can use the Delphi Method for giving weights to the short-listed critical factors.

The panel members should give weights to each factor as well as their reasoning. In this way, other panel members can evaluate the weights based on the reasons given and accept, modify, or reject those reasons and weights. For example, let us assume that after the application of Tier 1 by the transit agency, there still remain two or more project delivery methods as viable options. The agency has identified the following four factors to be considered in Tier 2:

- Shortening the schedule.

- Agency control over the project.

- Project cost.

- Competition among contractors.

The facilitator should ask the panel members to weigh each of the factors while giving their reasons for the weights selected. The facilitator can then use the collected weights and viewpoints and establish a weight for each factor. One possible approach could be to calculate an average weight for each factor based on responses. If there are large divergences in some responses, the facilitator should study that and comment on those cases. The outcome of this analysis should be distributed to the panel again for further consideration and modification. The facilitator will decide when to stop the process of the level of consensus desired.

It might be that our panel of experts provide the weights as results for c_1, c_2, c_3, and c_4 as 0.020, 0.15, 0.40, and 0.25.

> The rank order centroid method is a simple way of giving weight to a number of items ranked according to their importance. The decision-makers usually can rank items much more easily than give weight to them. This method takes those ranks as inputs and converts them to weights for each of the items.

1. List objectives in order from most important to least important.
2. Use the formulas for assigning weights.

The conversion is based on the following formula:

$$w_i = \left(\frac{1}{M}\right)\sum_{n=i}^{M}\frac{1}{n}$$

where M is the number of items and W_i is the weight for the i item. For example, if there are 4 items, the item ranked first will be weighted $(1 + 1/2 + 1/3 + 1/4)/4 = 0.52$, the second will be weighted $(1/2 + 1/3 + 1/4)/4 = 0.27$, the third $(1/3 + 1/4)/4 = 0.15$, and the last $(1/4)/4 = 0.06$. As shown in this example, the ROC is simple and easy to follow, but it gives weights which are highly dispersed (Fox 2018). As an example, consider the same factors to be weighted (shortening schedule, agency control over the project, project cost, and competition).

If they are ranked on the basis of their importance and influence on decision as

1. shortening schedule,

2. project cost,

3. agency control over the project, and

4. competition,

their weights would be 0.52, 0.27, 0.15, and 0.06, respectively. These weights almost eliminate the effect of the fourth factor, i.e., among competitors. This could be an issue for a decision-maker. For any four criteria that are ranked order will produce these weights.

10.2.1 Ratio Method for Weights

The Ratio Method is another simple way of calculating weights for a number of critical factors. A decision-maker should first rank all the items according to their importance. The next step is giving weight to each item based on its rank. The lowest ranked item will be given a weight of 10. The weight of the rest of the items should be assigned as multiples of 10. The last step is normalizing these raw weights. This process is shown in the example below. Note that the weights should not necessarily jump 10 points from one item to the next. Any increase in the weight is based on the subjective judgment of the decision-maker and reflects the difference between the importance of the items. Ranking the items in the first step helps in assigning more accurate weights (Table 10.1).

Normalized weights are simply calculated by dividing the raw weight of each item by the sum of the weights for all items. For example, normalized weight for the first item (shortening schedule) is calculated as 50/(50 + 40 + 20 + 10) = 41.7%. The sum of normalized weights is equal to 100% (41.7 + 33.3 + 16.7 + 8.3 = 100). Again, any four criteria that are ranked through order by importance will produce these weights.

10.2.2 Pairwise Comparison by Saaty (AHP)

In the pairwise comparison method, the decision-maker will compare each item with the rest of the group and give a preferential level to the item in each pairwise comparison. For example, if the item at hand is as important as the second one, the preferential level would be 1. If it is much more important, its level would be 10. After conducting all of the comparisons and determining the preferential levels, the numbers will be added up and normalized. The results are the weights for each item. Table 10.2 can be used as a guide for giving a preferential level score to an item while comparing it with another one. The following example shows the application of the pairwise comparison procedure. Referring to the four critical factors identified above, let us assume that shortening the schedule, project cost, and agency control of the project are the most important parameters in the project delivery selection decision. Following the pairwise comparison, the decision-maker should pick one of these factors (e.g., shortening the schedule) and compare it with the remaining factors and give a preferential level to it. For example, shortening the schedule is more important

TABLE 10.1 Ratio Method

Task/Item	Shorten Schedule	Project Cost	Agency Control	Competition
Ranking	1	2	3	4
Weighting	50	40	20	10
Normalizing	41.70%	33.30%	16.70%	8.30%

TABLE 10.2   Saaty's Nine-Point Scale

Intensity of Importance in Pairwise Comparisons	Definition
1	Equal importance
3	Moderate importance
5	Strong importance
7	Very strong importance
9	Extreme importance
2, 4, 6, 8	For comparing between the above
Reciprocals of above	In comparison of elements i and j, if i is 3 compared to j, then j is 1/3 compared to i

than project cost; in this case, it will be given a level of importance of the 5. Pairwise comparisons use the information in Table 10.2.

The decision-maker should continue the pairwise comparison and give weights to each factor. The weights, which are based on the preferential levels given in each pairwise comparison, should be consistent to the extent possible. The consistency is measured on the basis of the matrix of preferential levels.

Table 10.3 shows the rest of the hypothetical weights and the normalizing process, the last step in the pairwise comparison approach. Note that Column (5) is simply the sum of the values in Columns (1) through (4). Also note that if the preferential level of factor i to factor j is n, then the preferential level of factor j to factor i is simply $1/n$. The weights calculated for this exercise are 0.6, 0.1, 0.2, and 0.1 which add up to 1.0. Note that it is possible for two factors to have the same importance and weight.

Weights need to pass the consistency tests. Let λ be the largest eigenvalue of the pairwise comparison matrix.

Saaty proved that for consistent reciprocal matrix, the largest eigenvalue is equal to the size of comparison matrix, or $\lambda_{max} = n$. Then he gave a measure of consistency, called Consistency Index, as deviation or degree of consistency using the following formula:

$$CI = \frac{\lambda_{max} - n}{n-1}$$

For example, assume we have $\lambda_{max} = 3.0967$, and the size of comparison matrix is $n = 3$, thus, the consistency index is:

$$CI = \frac{\lambda_{max} - n}{n-1} = \frac{3.0967 - 3}{2} = 0.0484$$

Knowing the Consistency Index, the next question is how do we use this index. Again, Saaty proposed that we use this index by comparing it with the appropriate one. The appropriate Consistency Index is called Random Consistency Index (RI).

He randomly generated reciprocal matrix using scale $\frac{1}{9}, \frac{1}{8}, \ldots, 1, \ldots,$ 8, 9 and got the Random Consistency Index to see if it is about 10% or less. The average Random Consistency Index of sample size 500 matrices is shown in Table 10.3.

Now we will want to look at the Consistency Ratio, which is a comparison between Consistency Index and Random Consistency Index, or in formula:

$$CR = \frac{CI}{RI}$$

If the value of Consistency Ratio is smaller or equal to 10%, the inconsistency is acceptable. If the Consistency Ratio is greater than 10%, we need to revise the subjective judgment.

For our previous example, we have $CI = 0.0484$ and RI for $n = 3$ is **0.58**, then we have

$$CR = \frac{CI}{RI} = \frac{0\_0484}{0\_58} = 8\_3\% < 10\%$$

Thus, we conclude that our subjective evaluation preference is consistent.

We need to calculate this for each pairwise matrix (Table 10.4). The important condition is that if the matrix has a CR more than 0.10, then we must go back and revise the pairwise matrix until CR is less than 0.10.

10.2.3 Entropy Method

Another weighting method uses the real data and is called entropy. It has been shown to be effective in obtaining attribute weights (Fox 2020).

TABLE 10.3   Random Consistency Index (*RI*)

1	2	3	4	5	6	7	8	9	10
0	0	0.58	0.9	1.12	1.24	1.32	1.41	1.45	1.49

TABLE 10.4   Pairwise Comparison Example

	Shorten the Schedule	Project Cost	Agency Control	Competition	Total	Weights
	−1	−2	−3	−4	−5	−6
Shorten the schedule	1	5	2-May	8	16.5	16.5/27.225 = 0.60
Project cost	5-Jan	1	1/2	1	2.7	2.7/27/225 = 0.10
Agency control	5-Feb	2	1	2	5.4	5.4/27/225 = 0.20
Competition	8-Jan	1	1/2	1	2.625	2.625/27/225 = 0.10
				Total=	27.225	1

Entropy is the measure of uncertainty in the information using probability methods. It indicates that a broad distribution represents more uncertainty than does a sharply peaked distribution.

To determine the weights by the entropy method, the normalized decision matrix we call R_{ij} is considered. The equation used is

$$e_j = -k \sum_{i=1}^{n} R_{ij} \ln\left(R_{ij}\right)$$

Where $k = 1/ln(n)$ is a constant that guarantees that $0 \le e_j \le 1$. The value of n refers to the number of alternatives. The degree of divergence (d_j) of the average information contained by each attribute can be calculated as:

$$d_j = 1 - e_j$$

The more divergent the performance rating R_{ij}, for all i & j, then the higher the corresponding d_j and the more important the attribute B_j is. The weights are found by the equation,

$$w_j = \frac{\left(1 - e_j\right)}{\Sigma\left(1 - e_j\right)}.$$

We will look at a classic problem that many new car buyers face, and that is the best car to purchase. Table 10.5 provides common car performance data for several generic cars, a1–a6. Example 10.1 will walk you through obtaining entropy weights.

10.2.2.1 Example 10.1: Car Performance Data

(a) The data is given in Table 10.5.

TABLE 10.5   Car Data for Example 10.1

	Cost	Safety	Reliability	Performance	MPG City	MPG HW	Interior/ Style
a1	27.80	9.4	3	7.5	44	40	8.70
a2	28.50	9.6	4	8.4	47	47	8.10
a3	38.67	9.6	3	8.2	35	40	6.30
a4	25.50	9.4	5	7.8	43	39	7.50
a5	27.50	9.6	5	7.6	36	40	8.30
a6	36.20	9.4	3	8.1	40	40	8.00

(b) Sum the columns (as in Table 10.6).

TABLE 10. 6   Sum of Car Data for Example 10.1

	Cost	Safety	Reliability	Performance	MPG City	MPG HW	Interior/ Style
Sums	184.17	57.00	23.00	47.60	245.00	246.00	46.90

(c) Normalize the data (as in Table 10.7). Divide each data element in a column by the sum of the column.

TABLE 10.7   Normalized Car Data for Example 10.1

	Cost	Safety	Reliability	Performance	MPG City	MPG HW	Interior/ Style
a1	0.15	0.16	0.13	0.16	0.18	0.16	0.19
a2	0.15	0.17	0.17	0.18	0.19	0.19	0.17
a3	0.21	0.17	0.13	0.17	0.14	0.16	0.13
a4	0.14	0.16	0.22	0.16	0.18	0.16	0.16
a5	0.15	0.17	0.22	0.16	0.15	0.16	0.18
a6	0.20	0.16	0.13	0.17	0.16	0.16	0.17

(d) Use the entropy formula (as in Table 10.8), in the case $k = 6$.

$$e_j = -k \sum_{i=1}^{n} R_{ij} \ln(R_{ij})$$

TABLE 10.8   Entropy for Car Data in Example 10.1

	Cost	Safety	Reliability	Performance	MPG City	MPG HW	Interior/ Style
a1	−0.29	−0.30	−0.27	−0.29	−0.31	−0.30	−0.31
a2	−0.29	−0.30	−0.30	−0.31	−0.32	−0.32	−0.30
a3	−0.33	−0.30	−0.27	−0.30	−0.28	−0.30	−0.30
a4	−0.27	−0.30	−0.33	−0.30	−0.31	−0.29	−0.29
a5	−0.28	−0.30	−0.33	−0.29	−0.28	−0.30	−0.31
a6	−0.32	−0.30	−0.27	−0.30	−0.30	−0.30	−0.30

(e) Find e_j (as in Table 10.9).

TABLE 10.9 e_j for Car Data in Example 10.1

	Cost	Safety	Reliability	Performance	MPG City	MPG HW	Interior/ Style
e_j	0.993	0.999	0.985	0.999	0.997	0.999	0.997

(f) Compute weights by formula (as given in Table 10.10)

TABLE 10.10   Computed Weights for Car Data in Example 10.1

	Cost	Safety	Reliability	Performance	MPG City	MPG HW	Interior/ Style
w	0.226	0.032	0.484	0.032	0.097	0.032	0.097

(g) Check that weights sum to 1, as they did above.

(h) Interpret weights and rankings – reliability appears to be the most important criteria.

(i) Use these weights in further analysis.

10.3 SIMPLE ADDITIVE WEIGHTS (SAW) METHOD

The simple additive weights method is also called the weighted sum method (Fishburn 1967) and is the simplest and still one of the widest used of the MADM methods. Its simplistic approach makes it easy to use. Depending on the type of relational data used, we might either want the larger average or the smaller average.

10.3.1 Methodology

Here, each criterion (attribute) is given a weight, and the sum of all weights must be equal to 1. Each alternative is assessed with regard to every criterion (attribute). The overall or composite performance score of an alternative, with m criteria is,

$$P_i = \left(\sum\nolimits_{j=1}^{m} w_j m_{ij} \right) / m$$

Previously, it was argued that SAW should be used only when the decision criteria can be expressed in identical units of measure (*e.g., only dollars, only pounds, only seconds*). However, if all the elements of the decision table are normalized, then this procedure can be used for any type and any number of criteria. In that case, the overall performance score will take the following form,

$$P_i = \left(\sum\nolimits_{j=1}^{m} w_j m_{ijNormalized} \right) / m$$

where ($m_{ijNormalized}$) represents the normalized value of m_{ij}, and P_i is the overall or composite score of the alternative A_i. The alternative with the highest value of P_i is considered the best alternative.

10.3.2 Strengths and Limitations

The strengths are the ease of use and the normalized data allowing for comparison across many differing criteria. A limitation is that larger values mean larger is better and smaller values means smaller is better even though those meaning are not necessarily true. For example we would want to minimize costs, which is usually large. There is no the flexibility in this method to state which criterion should be larger or smaller to achieve better performance. This makes gathering useful data of the same relational value scheme (larger or smaller) essential.

10.3.3 Sensitivity Analysis

Sensitivity analysis should be applied to the weighing scheme employed to determine how sensitive the model is to the weights. Weighing can be arbitrary for a decision-maker or in order to obtain weights, you might choose to use a scheme to perform pairwise comparison as we show in AHP that we discuss later. Whenever subjectivity enters into the process for finding weights, sensitivity analysis is recommended. Please see later sections for a suggested scheme for dealing with sensitivity analysis for individual criteria weights.

10.3.4 Illustrative Examples of SAW

10.3.4.1 Example 10.2: Manufacturing Example

We have data (Table 10.11) for our three alternatives over four criteria.

We want to normalize the data in Table 10.11 for the alternatives. We normalize by first summing the data in each column and then dividing each data element by its column sum (Figure 10.1).

TABLE 10.11 Manufacturing Weights and Alternatives for Example 10.2

	C1	C2	C3	C4
Alts./Weights	0.2	0.15	0.4	0.25
A1	25	20	15	30
A2	10	30	20	30
A3	30	10	30	10

Alts / Weights	c1	c2	c3	c4
	0.2	0.15	0.4	0.25
a1	25	20	15	30
a2	10	30	20	30
a3	30	10	30	10
Total	65	60	65	70

	c1	c2	c3	c4
Normalized	0.2	0.15	0.4	0.25
a1	0.3846	0.3333	0.2308	0.4286
a2	0.1538	0.5000	0.3077	0.4286
a3	0.4615	0.1667	0.4615	0.1429

					Sum	Rank
a1	0.0769	0.0500	0.0923	0.1071	0.3264	2
a2	0.0592	0.0750	0.1231	0.1071	0.3360	1
a3	0.0710	0.0250	0.1846	0.0357	0.3376	3

FIGURE 10.1 Excel Screenshot of Simple Additives Weight Method for Example 10.2.

Our company has ranked order alternative 3 first, followed by alternative 2, then alternative 1.

10.3.4.2 Example 10.3: Updated Car Selection Problem (Data from Consumer's Report and US News and World Report Online Data)

We are considering six cars: Ford Fusion, Toyota Prius, Toyota Camry, Nissan Leaf, Chevy Volt, and Hyundai Sonata. For each car, we have data on seven criteria that were extracted from Consumer's Report and US News and World Report data sources. They are *cost, mpg city, mpg highway, performance, interior & style, safety,* and *reliability*. We provide the extracted information in Table 10.12.

Initially, we might assume that all weights are equal to obtain a baseline ranking. We substitute the rank orders (first to sixth) for the actual data. We compute the average rank attempting to find the best ranking (smaller is better). We find that our rank ordering is Fusion, Sonata, Camry, Prius, Volt, and Leaf.

We use pairwise comparisons to obtain weights. The weights are as follows:

Cost (0.3112)

MPG City (0.1336)

MPG Highway (0.0958)

Performance (0.0551)

Interior & Style (0.0499)

Safety (0.1294)

Reliability (0.2250)

TABLE 10.12   Raw Performance Data for Example 10.3

Cars	Cost ($000)	MPG City	MPG HW	Performance	Interior & Style	Safety	Reliability
Prius	27.80	44	40	7.5	8.7	9.4	3
Fusion	28.50	47	47	8.4	8.1	9.6	4
Volt	38.67	35	40	8.2	6.3	9.6	3
Camry	25.50	43	39	7.8	7.5	9.4	5
Sonata	27.50	36	40	7.6	8.3	9.6	5
Leaf	36.20	40	40	8.1	8.0	9.4	3

TABLE 10.13   Expected Values Car Data in Example 10.3

	Expected Values	Ranks
Prius	0.164	4
Fusion	0.178	3
Volt	0.142	6
Camry	0.187	1
Sonata	0.180	2
Leaf	0.149	5

We compute the expected value E[X] for each alternative (Table 10.13).

Based on this information we choose the Camry to be as our best alternative since it has the largest expected value.

10.4 TECHNIQUE OF ORDER PREFERENCE BY SIMILARITY TO THE IDEAL SOLUTION (TOPSIS)

The principle behind TOPSIS is simple: The chosen alternative should be as close to the ideal solution as possible and as far from the negative-ideal solution as possible. The ideal solution is formed as a composite of the best performance values exhibited (in the decision matrix) by any alternative for each attribute. The negative-ideal solution is the composite of the worst performance values. Proximity to each of these performance poles is measured in the Euclidean sense (the square root of the sum of the squared distances along each axis in the "attribute space").

10.4.1 Introduction

TOPSIS was developed in 1980 by K. Yoon and H. Ching-Lai at Kansas State University. Their premise was that the alternative chosen should have the shortest distance from the ideal solution and the farthest distance from the negative-ideal solution. In the late 1980s, the Department of Defense was using the TOPSIS program to rank order defense acquisitions across all branches for the POM and the budget cycle.

To begin with, assume we have m alternatives and n attributes. Further assume that either the decision-maker provides attributes weights or that we use something like Saaty's Analytical Hierarchy Process (AHP) to find the decision weights as applied to the attributes.

There are three key assumptions:

(1) Each attribute in the decision matrix takes on either monotonically increasing or monotonically decreasing utility.

(2) Weights for the attributes are either provided or can solved for with a procedure such as AHP.

(3) Any outcomes which is expressed in a nonnumerical way should be quantified through some appropriate scaling technique. Again, Saaty's 9-point scale works well.

10.4.2 Steps in TOPSIS

1. Create the $m \times n$ matrix with alternatives as rows and attributes as columns using the appropriate scaling techniques.

2. Transform to a normalized decision matrix which provided a nondimensional value and allows comparison across attributes.

3. Multiply the decision weights for the attributes by each value in the column.

4. Determine for each column both the maximum and minimum values.

5. Calculate the distance measures.

6. Calculate the relative closeness metric to the ideal solution.

7. Rank order the results in descending order.

10.4.3 TOPSIS Methodology

The TOPSIS process is carried out as follows:

Step 1.

Create an evaluation matrix consisting of m alternatives and n criteria, with the intersection of each alternative and criteria given as x_{ij}, giving us a matrix $(X_{ij})_{mxn}$ (Figure 10.2)

Step 2.

The matrix shown as D above is then normalized to form the matrix $R = (R_{ij})_{mxn}$, using the normalization method for $i = 1, 2, \ldots, m; j = 1, 2, \ldots n$

$$r_{ij} = \frac{x_{ij}}{\sqrt{\sum x_{ij}^2}}$$

Step 3.

Calculate the weighted normalized decision matrix. First we need the weights. Weights can come from either the decision-maker or by computation.

$$
D = \begin{array}{c} \\ A_1 \\ A_2 \\ A_3 \\ \\ \\ \\ \\ A_m \end{array}
\begin{array}{cccccc}
x_1 & x_2 & x_3 & & & x_n \\
\left[\begin{array}{cccccc}
x_{11} & x_{12} & x_{13} & \cdot & \cdot & x_{1n} \\
x_{21} & x_{22} & x_{23} & \cdot & \cdot & x_{2n} \\
x_{31} & x_{32} & x_{33} & \cdot & \cdot & x_{3n} \\
\cdot & \cdot & \cdot & \cdot & & \cdot \\
\cdot & \cdot & \cdot & \cdot & & \cdot \\
\cdot & \cdot & \cdot & \cdot & & \cdot \\
x_{m1} & x_{m2} & x_{m3} & \cdot & \cdot & x_{mn}
\end{array}\right]
\end{array}
$$

FIGURE 10.2 TOPSIS Evaluation Matrix.

Step 3a.

Use either the decision-maker's weights for the attributes $x_1, x_2, \ldots x_n$ or compute the weights through the use of Saaty's (1980) AHP's decision-maker weights method to obtain the weights as the eigenvector to the attributes versus attribute pairwise comparison matrix.

$$
\sum_{j=1}^{n} w_j = 1
$$

The sum of the weights over all attributes must equal 1 regardless of the method used.

Step 3b.

Multiply the weights to each of the column entries in the matrix from *Step 2* to obtain the matrix, *T*.

$$
T = (t_{ij})_{mxn} = (w_j r_{ij})_{mxn}, i = 1, 2, \ldots, m
$$

Step 4.

Determine the worst alternative (A_w) and the best alternative (A_b): Examine each attribute's column and select the largest and smallest values appropriately. If the values imply larger is better (profit), then the best alternatives are the largest values, and if the values imply smaller is better (such as cost), then the best alternative is the smallest value.

$$A_w = \left\{ \langle \max(t_{ij}|i=1,2,\ldots,m \mid j \in J_-), \langle \min(t_{ij}|i=1,2,\ldots,m) \mid j \in J_+ \rangle \right\}$$
$$\equiv \left\{ t_{wj}|j=1,2,\ldots,n \right\},$$

$$A_{wb} = \left\{ \langle \min(t_{ij}|i=1,2,\ldots,m \mid j \in J_-), \langle \max(t_{ij}|i=1,2,\ldots,m) \mid j \in J_+ \rangle \right\}$$
$$\equiv \left\{ t_{bj}|j=1,2,\ldots,n \right\},$$

where
$J_+ = \{j=1,2,\ldots n|j)$ associated with the criteria having a positive impact,
and
$J_- = \{j=1,2,\ldots n|j)$ associated with the criteria having a negative impact.

We suggest that, if possible, make all entry values in terms of positive impacts.

Step 5.

Calculate the L2-norm distance between the target alternative i and the worst condition A_w, and the distance between the alternative i and the best condition A_b

$$d_{iw} = \sqrt{\sum_{j=1}^{n}(t_{ij}-t_{wj})^2}, \, i = 1,2,\ldots m$$

$$d_{ib} = \sqrt{\sum_{j=1}^{n}(t_{ij}-t_{bj})^2}, \, i = 1,2,\ldots m$$

where d_{iw} and d_{ib} are L2-norm distances from the target alternative i to the worst and best conditions, respectively.

Step 6.

Calculate the similarity to the worst condition:

$$s_{iw} = \frac{d_{iw}}{(d_{iw}+d_{ib})}, 0 \le s_{iw} \le 1, i = 1,2,\ldots,m$$

$.S_{iw} = 1$ if and only if the alternative solution has the worst condition; and $S_{iw} = 0$ if and only if the alternative solution has the best condition.

Step 7.
Rank the alternatives according to their value from S_{iw} $(i = 1, 2, \ldots, m)$.

10.4.4 Normalization

Two methods of normalization that have been used to deal with incongruous criteria dimensions are linear normalization and vector normalization. Linear normalization can be calculated as in *Step 2* of the TOPSIS process. Vector normalization was incorporated with the original development of the TOPSIS method (Hwang and Yoon 1981) and is calculated using the following formula:

$$r_{ij} = \frac{x_{ij}}{\sqrt{\sum x_{ij}^2}}$$

for $i = 1, 2 \ldots, m; j = 1, 2, \ldots n.$

10.4.4.1 Example 10.4: Selecting a New Car with AHP

We want to choose among four cars: Civic, Saturn, Ford, and Mazda. We considered four criteria of body style, reliability, fuel economy, and cost.

Step 1. Obtain the decision weights. Choose any method. We use the AHP method to obtain the weights as 0.0549, 0.4145, 0.4086, and 0.1219 as shown in Figure 10.3.

Step 2. We create a matrix of the alternatives rated in each criterion. We used the 1–9 scale as before (Figure 10.4).

Step 3. Square the entries, sum, and take square root to the totals (Figure 10.5).

Step 4. Normalize the entries in Step 2 by dividing them by the square root of the column total (Figure 10.6).

Step 5. Multiply all entries in the normalized matrix by the criterion weights found in step 1 (Figure 10.7).

Step 6. Pick the largest and smallest value in each column. We use max(column values) and min(column values) to obtain (as given in Figure 10.8).

Step 7. Compute the separation from the max ideal (as in Figure 10.9).
(Value-maximum value)².

Step 8. Compute the separation from the min idea 1 (as given in Figure 10.10).
(Value-minimum value)².

Step 9. Calculate (Figure 10.11) the relative closeness to the ideal solution $C_i^* = S'_i/(S_i^* + S'_i)$

Style	Reliability	Fuel Economy	Cost
1	0.1667	0.1111	0.3333
6	1.0000	1.0000	4.0000
9	1.0000	1.0000	3.0000
3	0.2500	0.3333	1.0000
19	2.4167	2.4444	8.3333

0.0526	0.0690	0.0455	0.04
0.3158	0.4138	0.4091	0.48
0.4737	0.4138	0.4091	0.36
0.1579	0.1034	0.1364	0.12

Style	Reliability	Fuel Economy	Cost
0	1	0	0
0.0690	0.4138	0.4138	0.1034
0.0551	0.4119	0.4104	0.1225
0.0549	0.4146	0.4086	0.1220
0.0549	0.4146	0.4086	0.1219
0.0549	0.4146	0.4086	0.1219
0.0549	0.4146	0.4086	0.1219
0.0549	0.4146	0.4086	0.1219
0.0549	0.4146	0.4086	0.1219
0.0549	0.4146	0.4086	0.1219
0.0549	0.4146	0.4086	0.1219
0.0549	0.4146	0.4086	0.1219

FIGURE 10.3   Excel Screenshot of Step 1 of AHP Method.

Cars	Criterion			
	Style	Reliability	Fuel Economy	Cost
Civic	7	9	9	8
Saturn	8	7	8	7
Ford	9	6	8	9
Mazda	6	7	8	6

FIGURE 10.4   Excel Screenshot of Step 2 of AHP Method.

49	81	81	64
64	49	64	49
81	36	64	81
36	49	64	36
230	215	273	230
15.16575089	14.6628783	16.5227116	15.16575089

FIGURE 10.5   Excel Screenshot of Step 3 of AHP Method.

0.461566331	0.613794906	0.54470478	0.527504379
0.527504379	0.477396038	0.48418203	0.461566331
0.593442426	0.409196604	0.48418203	0.593442426
0.395628284	0.477396038	0.48418203	0.395628284

FIGURE 10.6   Excel Screenshot of Step 4 of AHP Method.

Multiply by weights.			
Assume weights are			
0.054930536	0.414559405	0.40860391	0.121906151
0.025354086	0.254454451	0.2225685	0.064306028
0.028976098	0.197909017	0.19783867	0.056267775
0.03259811	0.1696363	0.19783867	0.072344282
0.021732074	0.197909017	0.19783867	0.048229521

FIGURE 10.7   Excel Screenshot of Step 5 of AHP Method.

Pick the largest and smallest in each column			
0.03259811	0.254454451	0.2225685	0.072344282
0.021732074	0.1696363	0.19783867	0.048229521

FIGURE 10.8   Excel Screenshot of Step 6 of AHP Method.

Separation from max ideal.

5.24759E-05	0	0	0.000258454	0.000311	0.017633
1.3119E-05	0.003197386	0.00061156	6.46135E-05	0.003887	0.062343
0	0.007194119	0.00061156	0.000581522	0.008387	0.091582
0.000118071	0.003197386	0.00061156	0	0.003927	0.062666

FIGURE 10.9   Excel Screenshot of Step 7 of AHP Method.

Separation from minimum ideal

1.3119E-05	0.007194119	0.00061156	6.46135E-05	0.007883	0.088789
5.24759E-05	0.000799347	0	0.000258454	0.00111	0.033321
0.000118071	0	0	0	0.000118	0.010866
0	0.000799347	0	0.000581522	0.001381	0.03716

FIGURE 10.10   Excel Screenshot of Step 8 of AHP Method.

Relative Closeness	
	0.834308335
	0.348310581
	0.106064211
	0.372248112

FIGURE 10.11   Excel Screenshot of Step 9 of AHP Method.

This results in the following relative closeness values for our cars.

Civic: 0.8343

Saturn: 0.3483

Ford: 0.1066

Mazda: 0.3722

Based on this information, our ranking has the first position occupied by Civic, followed by the Mazda, the Saturn, and the Ford.

10.5 MODELING OF RANKING UNITS USING DATA ENVELOPMENT ANALYSIS (DEA) WITH LINEAR PROGRAMMING

Data envelopment analysis (DEA), occasionally called frontier analysis, was first put forward by Charnes, Cooper, and Rhodes in 1978. It is a performance measurement technique which, as we shall see, can be used for evaluating the *relative efficiency* of *decision-making units* (*DMUs*) in organizations. Here, a DMU is a distinct unit within an organization that has flexibility with respect to some of the decisions it makes, but not necessarily complete freedom with respect to these decisions.

Examples of such units to which DEA has been applied are banks, police stations, hospitals, tax offices, prisons, defense bases (army, navy, air force), schools, and university departments. Note here that one advantage of DEA is that it can be applied to nonprofit making organizations.

Since the technique was first proposed, much theoretical and empirical work has been done. Many studies have been published dealing with applying DEA in real-world situations. Obviously, there are many more unpublished studies, e.g., done internally by companies or by external consultants.

We will initially illustrate DEA by means of a small example. More about DEA can be found online using Google: "Data Envelopment Analysis". Note here that much of what you will see below is a graphical (pictorial) approach to DEA. This is very useful if you are attempting to explain DEA to those less technically qualified. There is a mathematical approach to DEA that can be adopted, however. We will present the single measure first to demonstrate the idea and then move to multiple measures and use linear programming methodology from our course.

10.5.1 Example 10.5: Banks

Consider several bank branches. For each branch, we have a single output measure (number of personal transactions completed) and a single input measure (number of staff).

The data we have is as shown in Table 10.13.

For example, for the Branch 2 in one year, there were 44,000 transactions relating to personal accounts, and 16 staff members were employed.

How then can we compare these branches and measure their performance using this data?

TABLE 10.13 Bank Transactions and Staff for Example 10.5

Branch	Personal Transactions ('000s)	Number of Staff
1	125	18
2	44	16
3	80	17
4	23	11

TABLE 10.14 Personal Transactions for Example 10.5

Branch	Personal Transactions ('000s)
1	6.94
2	2.75
3	4.71
4	2.09

A commonly used method is *ratios*. Typically, we take some output measure and divide it by some input measure. Note the terminology here, we view branches as taking *inputs* and converting them (with varying degrees of efficiency, as we shall see below) into *outputs*.

For our bank branch example, we have a single input measure, the number of staff, and a single output measure, the number of personal transactions. Therefore, we have the data as shown in Table 10.14.

Here, we can see that Branch 1 has the highest ratio of personal transactions per staff member, whereas Branch 4 has the lowest ratio of personal transactions per staff member.

As Branch 1 has the highest ratio of 6.94, we can compare all other branches to it and calculate their *relative efficiency* with respect to Branch 1. To do this, we divide the ratio for any branch by 6.94 (the value for Croydon) and multiply by 100 to convert to a percentage. This gives data (as shown in Table 10.15).

The other branches do not compare well with Branch 1, so are presumably performing less well. That is, they are relatively less efficient at using their given input resource (staff members) to produce output (number of personal transactions).

We could, if we wish, use this comparison with Branch 1 to set *targets* for the other branches.

For example, we could set a target for Branch 4 of continuing to process the same level of output but with one less member of staff. This is an example of an **input target** as it deals with an input measure.

TABLE 10.15   Relative Efficiency for Example 10.5

Branch	Relative Efficiency
1	$100 \ (6.94/6.94) = 100\%$
2	$100 \ (2.75/6.94) = 40\%$
3	$100 \ (4.71/6.94) = 68\%$
4	$100 \ (2.09/6.94) = 30\%$

An example of an ***output target*** would be for Branch 4 to increase the number of personal transactions by 10% (e.g., by obtaining new accounts).

Plainly, in practice, we might well set a branch a mix of input and output targets which we want it to achieve. We can use linear programming.

10.6 ILLUSTRATIVE EXAMPLES

10.6.1 Example 10.6: Bank Example with Linear Programming

Typically, we have more than one input and one output. For the bank branch example, suppose now that we have two output measures (number of personal transactions completed and number of business transactions completed) and the same single input measure (number of staff) as before.

The data we have is as shown in Table 10.16.

We start by scaling (via ratios) the inputs and outputs to reflect the ratio of 1 unit (Table 10.17).

Pick a DMU to maximize: *E1*, *E2*, *E3*, or *E4*

Let *W1* and *W2* be the personal and business transactions at each branch. In this example, we choose to maximize branch one, *E1*.

The LP formulation is:

Objective FunctionL Maximize *E1*

Subject to:

$$E1 = 6.94 \ W1 + 2.78 \ W2$$

$$E2 = 2.75 \ W1 + 1.25 \ W2$$

$$E3 = 4.71 \ W1 + 3.24 \ W2$$

$$E4 = 2.09 \ W1 + 1.09 \ W2$$

$$E1 \le 1$$

$$E2 \le 1$$

TABLE 10.16 Bank Example 10.6 for Linear Programming

Branch	Personal Transactions ('000s)	Business Transactions ('000s)	Number of Staff
1	125	50	18
2	44	20	16
3	80	55	17
4	23	12	11

TABLE 10.17 Bank Example 10.6 for Scaling

Branch	Personal Transactions ('000s)	Business Transactions ('000s)	Number of Staff
1	125/18 = 6.94	50/18 = 2.78	18/18 = 1
2	44/64 = 2.75	20/16 = 1.25	16/16 = 1
3	80/17 = 4.71	55/17 = 3.24	17/17 = 1
4	23/11 = 2.09	12/11 = 1.09	11/11 = 1

$$E3 \leq 1$$

$$E4 \leq 1$$

Non-negativity

The Excel output is as shown in Figure 10.12.

The Excel Solution is included in Figure 10.13.

The sensitivity report from Excel is as shown in Figure 10.13.

Now, what did we learn from this? If we rank-ordered the branches on efficiency performance of our inputs and outputs, we find:

Branch 1: 100%

Branch 3: 100%

Branch 2: 43.2%

Branch 4: 36.2%

We know we need to improve on branch 2 and branch 4's performance while not losing our efficiency in branches 1 and 3. A better interpretation could be that the practices and procedures used by the other branches were to be adopted by Branch 4, then Branch 4 they could improve their performance.

maximize E1			E1	1
	0.431809		E2	0.431809
			E3	1
			E4	0.361777
			w1	0.04877
subject to			w2	0.238152
	1	1	w3	0
	0.431809	0.431809		
	1	1		
	0.361777	0.361777		
	1	1		
	0.431809	1		
	1	1		
	0.361777	1		

FIGURE 10.12A   Excel Screenshot of LP Formulation for Example 10.6.

Target Cell (Max)

Cell	Name	Original Value	Final Value
$M$4	w2	0	0.432

	Adjustable Cells		
Cell	**Name**	**Original Value**	**Final Value**
$P$3	E1	0	1
$P$4	E2	0	0.432
$P$5	E3	0	1.000
$P$6	E4	0	0.362
$P$7	w1	0	0.049
$P$8	w2	0	0.238
$P$9	w3	0	0

Constraints

Cell	Name	Cell Value	Formula	Status	Slack
$M$9		1.000	$M$9=$N$9	Not Binding	0.000
$M$10		0.432	$M$10=$N$10	Not Binding	0.000
$M$11		1.000	$M$11=$N$11	Not Binding	0.000
$M$12		0.362	$M$12=$N$12	Not Binding	0.000
$M$13		1.000	$M$13<=1	Binding	0.000
$M$14		0.432	$M$14<=1	Not Binding	0.568
$M$15		1.000	$M$15<=1	Binding	0.000
$M$16		0.362	$M$16<=1	Not Binding	0.638

FIGURE 10.12B   Excel Screenshot of LP Formulation Solution for Example 10.6.

Adjustable Cells

Cell	Name	Final Value	Reduced Cost	Objective Coefficient	Allowable Increase	Allowable Decrease
$P$3	E1	1.000	0.000	0.000	1.000E+30	0.321
$P$4	E2	0.432	0.000	1.000	1.000E+30	1.000
$P$5	E3	1.000	0.000	0.000	1.000E+30	0.110
$P$6	E4	0.362	0.000	0.000	1.000E+30	0.588
$P$7	w1	0.049	0.000	0.000	0.373	0.932
$P$8	w2	0.238	0.000	0.000	0.641	0.149
$P$9	w3	0.000	0.000	0.000	0.000	1.000E+30

Constraints

Cell	Name	Final Value	Shadow Price	Constraint R.H. Side	Allowable Increase	Allowable Decrease
$M$9		1.000	0.321	0.000	0.475	0.141
$M$10		0.432	-1.000	0.000	0.432	0.568
$M$11		1.000	0.110	0.000	0.165	0.322
$M$12		0.362	0.000	0.000	0.362	0.638
$M$13		1.000	0.321	1.000	0.475	0.141
$M$14		0.432	0.000	1.000	1.000E+30	0.568
$M$15		1.000	0.110	1.000	0.165	0.322
$M$16		0.362	0.000	1.000	1.000E+30	0.638

FIGURE 10.13 Excel Screenshot of LP Formulation Sensitivity Analysis for Example 10.6.

For any DMU_0, let X_i be the inputs and Y_i be the outputs. Let X_0 and Y_0 be the DMU being modeled.

$$Min\ \theta$$

Subject to:

$$\Sigma\lambda_i X_i \le \theta X_0$$

$$\Sigma\lambda_i Y_i \le Y_0$$

$$\lambda i \ge 0$$

Non-negativity

10.6.1.1 Example 10.7: DMU Manufacturing

Consider the following process (Trick 1996, 2014; Chapter 12; Winston 1995) where we have three DMUs where each have 2 inputs and 3 outputs as shown in Table 10.18.

TABLE 10.18   DMU Data for Example 10.7

DMU	Input#1	Input#2	Output#1	Output#2	Output#3
1	5	14	9	4	16
2	8	15	5	7	10
3	7	12	4	9	13

We define the following decision variables:

t_i = Value of a single unit of output of DMU i, for $i=1,2,3$

w_i = Cost or weights for one unit of inputs of DMU i, for $i=1,2$

efficiency$_i$ = (Total value of i's outputs)/(total cost of i's inputs), for $i = 1, 2, 3$

The following modeling assumptions are made:

a. No unit will have an efficiency more than 100%.

b. If any efficiency is less than 1, then it is inefficient.

c. We should scale the costs as the costs of the inputs equal 1 for each linear program. For example, we will use $5w_1 + 14w_2 = 1$ in our program for DMU_1.

d. All values and weights must be strictly positive, so we use a constant such as 0.0001 in lieu of 0.

To calculate the efficiency of unit 1, we define the linear program:

Objective function: Maximize $9t_1 + 4t_2 + 16t_3$

Subject to:

$$-9t_1 - 4t_2 - 16t_3 + 5w_1 + 14w_2 \geq 0$$

$$-5t_1 - 7t_2 - 10t_3 + 8w_1 + 15w_2 \geq 0$$

$$-4t_1 - 9t_2 - 13t_3 + 7w_1 + 12w_2 \geq 0$$

$$5w_1 + 14w_2 = 1$$

$$t_i \geq 0.0001, i = 1, 2, 3$$

$$w_i \geq 0.0001, i = 1, 2$$

Non-negativity

To calculate the efficiency of unit 2, we define the linear program as
Objective function: Maximize $5t_1 + 7t_2 + 10t_3$
Subject to:

$$-9t_1 - 4t_2 - 16t_3 + 5w_1 + 14w_2 \geq 0$$

$$-5t_1 - 7t_2 - 10t_3 + 8w_1 + 15w_2 \geq 0$$

$$-4t_1 - 9t_2 - 13t_3 + 7w_1 + 12w_2 \geq 0$$

$$8w_1 + 15w_2 = 1$$

$$t_i \geq 0.0001, i = 1, 2, 3$$

$$w_i \geq 0.0001, i = 1, 2$$

Non-negativity

To calculate the efficiency of unit 3, we define the linear program as:
Objective function: Maximize $4t_1 + 9t_2 + 13t_3$
Subject to:

$$-9t_1 - 4t_2 - 16t_3 + 5w_1 + 14w_2 \geq 0$$

$$-5t_1 - 7t_2 - 10t_3 + 8w_1 + 15w_2 \geq 0$$

$$-4t_1 - 9t_2 - 13t_3 + 7w_1 + 12w_2 \geq 0$$

$$7w_1 + 12w_2 = 1$$

$$t_i \geq 0.0001, i = 1, 2, 3$$

$$w_i \geq 0.0001, i = 1, 2$$

Non-negativity

The linear programming solutions show the efficiencies as:

$$u_1 = u_3 = i, u_2 = 0.77303.$$

Interpretation: u_2 is operating at 77.303% of the efficiency of u_1 and u_3. You could concentrate some improvements or best practices from u_1 or u_3 for u_2. An examination of the dual prices for the linear program of DMU_2 yields $\lambda_1 = 0.261538$, $\lambda_2 = 0$, and $\lambda_3 = 0.661538$. The average output vector for DMU_2 can be written as:

$$0.261538 \begin{bmatrix} 9 \\ 4 \\ 16 \end{bmatrix} + 0.661538 \begin{bmatrix} 4 \\ 9 \\ 13 \end{bmatrix} = \begin{bmatrix} 5 \\ 7 \\ 12.785 \end{bmatrix}$$

And the average input vector can be written as

$$0.261538 \begin{bmatrix} 5 \\ 14 \end{bmatrix} + 0.661538 \begin{bmatrix} 7 \\ 12 \end{bmatrix} = \begin{bmatrix} 5.938 \\ 11.6 \end{bmatrix}.$$

We may clearly see that the inefficiency is output 3 where $12.785 - 10 = 2.785$ more of output 3. This helps to focus on treating the inefficiency found for units.

Sensitivity analysis: This is also called "what if" analysis. So, let's assume that without management engaging some training for u_2 that u_2 production dips with output #2 dippings 9 units of output while the input hours increases from 15 to 16 hours.

We find that changes in technology coefficients are easily handled in resolving the LPs. Since u_2 is affected, we might only modify and solve the LP concerning u_2. We find with these changes that u_2 is now only 74% as effective as u_1 and u_3.

10.7 TECHNOLOGY FOR MULTI-ATTRIBUTE DECISION-MAKING (MADM)

As we discussed earlier, multi-attribute decision analysis is a common approach to help decision-makers identify the "best" choices out of a set of alternatives. However, there is rarely a a "best answer" to a problem, and it becomes necessary to create a ranking of alternatives from most to least effective, based on some sort of criterion. A significant difficulty is that the problem may originate from multiple sources and is sensitive to weightings. Technology, such as Excel or R, provides a nice format to help decision-makers rapidly explore the impact of weightings on the potential solution for MADM problems.

10.7.1 Technology and Simple Additive Weights

We discussed simple additive weights back in Section 10.3 and provided a car selection problem (Example 10.3) that addressed a common problem faced by many consumers: Which is the best car? In this section, we will look at using technology, specifically Excel, to solve the problem. We have created several user-friendly templates for SAW and TOPSIS that are useful for such weighting methods. Figures 10.14 and 10.15 provide an illustration of the SAW template, solving the car selection problem (Example 10.3) and will cover it in more detail in Chapter 11.

Sum of Additive Weight Method, SAW										Weight	3	
		Fill in yellow highlighted areas									1 Pairwsie	
											2 enter your own	
Enter the number of alternatives					9	Up to 20 alternatives					3 Entropy	
Enter then number of criteria					4	Up to 8 criteria					4 ROC	
											5 Ratio	
Enter the weights for your criteria below.					You may use the any weighting scheme on workbook: Weights 1-use AHP weights (go to Sheet Weights) or 2- enter your							
		C1	C2	C3	C4	C5	C6	C7	C8	sums		
Enter your weights		0	0	0	0	0	0	0	0	0		
Weights from any selected method		0.27046	0.27046	0.270460115	0.188619656	0	0	0	0	1		
Are you entering raw data or raw ranks? 1- for raw data, 2-raw ranks										1		
Raw data must be entered so that larger is better, ranking entered so that smaller is better.												
Enter the data or ranks below		Criteria										
		Cost	Safety	Reliability	Performance	MPG City	MPG HW	Interior/S	NA			
Alternatives		C1	C2	C3	C4	C5	C6	C7	C8			
	1	4	1	3	2	0	0	0	0			Prius
	2	3	2	4	3	0	0	0	0			Fusion
	3	3	3	2	2	0	0	0	0			Volt
	4	2	3	3	1	0	0	0	0			Camry
	5	2	4	2	2	0	0	0	0			Sonata
	6	2	2	1	3	0	0	0	0			Leaf
	7	3	1	3	2	0	0	0	0			
	8	1	2	2	4	0	0	0	0			
	9	1	3	1	3	0	0	0	0			
	10	0	0	0	0	0	0	0	0			
	11	0	0	0	0	0	0	0	0			
	12	0	0	0	0	0	0	0	0			
	13	0	0	0	0	0	0	0	0			
	14	0	0	0	0	0	0	0	0			
	15	0	0	0	0	0	0	0	0			
	16	0	0	0	0	0	0	0	0			
	17	0	0	0	0	0	0	0	0			
	18	0	0	0	0	0	0	0	0			
	19	0	0	0	0	0	0	0	0			
	20	0	0	0	0	0	0	0	0			
sums		21	21	21	22	0	0	0	0			

FIGURE 10.14 Screenshot of SAW Template.

Weights used		0.27046	0.27046	0.270460115	0.188619656	0	0	0	0	Sums of alternative	Rank	
	1	0.051516	0.012879	0.038637159	0.017147241	0	0	0	0	0.120179666	2	1
	2	0.038637	0.025758	0.051516212	0.025720862	0	0	0	0	0.14163234	1	2
	3	0.038637	0.038637	0.025758106	0.017147241	0	0	0	0	0.120179666	2	3
	4	0.025758	0.038637	0.038637159	0.008573621	0	0	0	0	0.111606045	5	4
	5	0.025758	0.051516	0.025758106	0.017147241	0	0	0	0	0.120179666	2	5
	6	0.025758	0.025758	0.012879053	0.025720862	0	0	0	0	0.090116128	8	6
	7	0.038637	0.012879	0.038637159	0.017147241	0	0	0	0	0.107300613	6	7
	8	0.012879	0.025758	0.025758106	0.034294483	0	0	0	0	0.098689748	7	8
	9	0.012879	0.038637	0.012879053	0.025720862	0	0	0	0	0.090116128	8	9

FIGURE 10.15 Screenshot of Results.

10.7.1.2 Example 10.7: Cars Using SAW

Next, we apply a scheme to the weights and still use the ranks 1–6 as before. For example, we apply a technique similar to the pairwise comparison that we discussed in the AHP section (Section 10.4). Using the pairwise comparison to obtain new weights, we obtain a new ordering:

Camry, Sonata, Fusion, Prius, Leaf, and Volt.

The changes in our results of the rank ordering differ from by changing the weights (we used equal weights to show the sensitivity that the model has to the given criteria weights. We assume that the criteria in order of importance are cost, reliability, MPG City, safety, MPG HW, performance, and interior & style.

10.7.1.3 Example 10.8: Car Selection Revisited Using TOPSIS

We might assume that our decision-maker weights from the AHP section (Section 10.4) are still valid for our use, so we can use the same PCM matrix as before. We must place the input data for the alternatives in the same order as the prioritized criteria. The input data look like this 8 × 8 matrix:

$$
AltM := \begin{bmatrix}
27.8 & 9.4 & 3 & 7.544 & 40 & 8.7 & 0 \\
28.5 & 9.6 & 4 & 8.447 & 47 & 8.1 & 0 \\
38.668 & 9.6 & 3 & 8.235 & 40 & 6.3 & 0 \\
25.5 & 9.4 & 5 & 7.843 & 39 & 7.5 & 0 \\
27.5 & 9.6 & 5 & 7.636 & 40 & 8.3 & 0 \\
36.2 & 9.4 & 3 & 8.140 & 40 & 8 & 0 \\
0 & 0 & 0 & 0 & 0 & 0 & 0 \\
0 & 0 & 0 & 0 & 0 & 0 & 0
\end{bmatrix}
$$

As before, our criteria weights are:

$$CR := 0.02149944074$$

$$DMWR := [0.361233126487974, 0.209324398363431,$$
$$0.144589995206757, 0.116672945567780, 0.0801478041928341,$$
$$0.0529870560310582, 0.0350446742839783, 0.]$$

and the CR is less than 0.1.

> *Part 1 weights* (*k, R, A, RA, AltM, PCM, Dweights, lsize*);
"CR and weights =", 0.02149944074, [0.361233126487974,
0.209324398363431, 0.144589995206757, 0.116672945567780,
0.0801478041928341, 0.0529870560310582,
0.0350446742839783, 0.]

The TOPSIS values for our alternatives are:

$$Ranks := \begin{bmatrix} 0.614967288202874 \\ 0.715258858184150 \\ 0.0614818245926484 \\ 0.908784041886248 \\ 0.810338906822239 \\ 0.182468079847254 \end{bmatrix}$$

Hence, the order of alternatives are:

Camry (0.9087), Sonata (0.8103), Fusion (0.7152), Prius (0.6149), Leaf (0.1824) and Volt (0.06148).

10.8 EXERCISES

In each problem, use SAW and then TOPSIS to find the ranking under these weighted conditions:

(a) All weights are equal.

(b) Choose and state your weights.

10.1. For a given hospital, rank order the procedure using the data in Table 10.19.

TABLE 10.19   Data for Exercise 10.1

	Procedure			
	1	2	3	4
Profit	$200	$150	$100	$80
X-Ray times	6	5	4	3
Laboratory Time	5	4	3	2

10.2. For a given hospital, rank order the procedure using the data in Table 10.20.

TABLE 10.20   Data for Exercise 10.2

	Procedure			
	1	2	3	4
Profit	$190	$150	$110	980
X-Ray times	6	5	5	3
Laboratory Time	5	4	3	3

10.3. Rank order the following threats in Table 10.21.

TABLE 10.21   Data for Exercise 10.3

Threat Alternatives\Criterion	Reliability of Threat Assessment	Approximate Associated Deaths (000)	Cost to Fix Damages in (Millions)	Location Density in Millions	Destructive Psychological Influence	Number of Intelligence-Related Tips
Dirty bomb threat	0.4	10	150	4.5	9	3
Anthrax-bio terror threat	0.45	0.8	10	3.2	7.5	12
DC-road and bridge network threat	0.35	0.005	300	0.85	6	8
NY subway threat	0.73	12	200	6.3	7	5
DC metro threat	0.69	11	200	2.5	7	5
Major bank robbery	0.81	0.0002	10	0.57	2	16
FAA threat	0.7	0.001	5	0.15	4.5	15

10.4. Consider a scenario where we want to move and rank the cities in Table 10.22.

TABLE 10.22   Data for Exercise 10.4

City	Affordability of Housing (Average Home Cost in Hundreds of Thousands)	Cultural Opportunities – Events Per Month	Crime rate – Number of Reported # Crimes Per Month (in Hundreds)	Quality of Schools on Average (Quality Rating Between [0, 1])
1	250	5	10	0.75
2	325	4	12	0.60
3	676	6	9	0.81
4	1,020	10	6	0.80
5	275	3	11	0.35
6	290	4	13	0.41
7	425	6	12	0.62
8	500	7	10	0.73
9	300	8	9	0.79

10.5. Consider rating departments at a college.

The following information is provided in Table 10.23.

TABLE 10.23   Data for Exercise 10.5

DMU Departments	Inputs # Faculty	Outputs Student Credit Hours	Outputs Number of Students	Outputs Total Degrees
Unit 1	25	18,341	9,086	63
Unit 2	15	8,190	4,049	23
Unit 3	10	2,857	1,255	31
Unit 4	33	22,277	6,102	31
Unit 5	12	6,830	2,910	19

Formulate and solve the DEA model and rank order the five departments.

10.6. Consider ranking companies within a larger organization. For simplification reasons, we will consider only six companies and their information in Table 10.24.

TABLE 10.24   Company Data for Exercise 10.6

Companies	Inputs # Size of Unit	Output #1	Output #2	Output #3
Unit 1	120	18,341	9,086	63
Unit 2	110	8,190	4,049	23
Unit 3	100	2,857	1,255	31
Unit 4	135	22,277	6,102	31
Unit 5	120	6,830	2,910	19
Unit 6	95	5,050	1835	12

10.7. Given the input–output data in Table 10.25 for three hospitals where inputs are the number of beds and labor hours in thousands per month, and outputs, all measured in hundreds, are patient-days for patients under 14, patient-days for patients between 14 and 65, and patient-days for patients over 65. Determine the efficiency of the three hospitals.

TABLE 10.25   Input/Output Data for Hospitals in Exercise 10.7

Hospital	Inputs		Outputs		
	1	2	1	2	3
1	5	14	9	4	16
2	8	15	5	7	10
3	7	12	4	9	13

10.8. Consider ranking 4 bank branches in a particular city. The inputs are:

Input 1 = Labor hours in hundred per month

Input 2 = Space used for tellers in hundreds of square feet

Input 3 = Supplies used in dollars per month

Output 1 = Loan applications per month

Output 2 = Deposits made in thousands of dollars per month

Output 3 = Checks processed thousands of dollars per month

The following data (Table 10.26) is for the bank branches.

TABLE 10.26   Bank Data for Exercise 10.8

Branches	Input 1	Input 2	Input 3	Output 1	Output 2	Output 3
1	15	20	50	200	15	35
2	14	23	51	220	18	45
3	16	19	51	210	17	20
4	13	18	49	199	21	35

REFERENCES AND SUGGESTED READINGS

Alinezhad, A. & Amini, A. (2011). Sensitivity analysis of TOPSIS technique: the results of change in the weight of one attribute on the final ranking of alternatives, *Journal of Optimization in Industrial Engineering*, 7(2011), 23–28.

Baker, T. & Zabinsky, Z. (2011). A multicriteria decision making model for reverse logistics using analytical hierarchy process, *Omega*, (39), 558–573.

Burden, R. & Faires, D. (2013). *Numerical analysis* (9th ed.). Cengage Publishers, Boston, MA.

Butler, J., Jia, J. & Dyer, J. (1997). Simulation techniques for the sensitivity analysis of multi-criteria decision models, *European Journal of Operations Research*, 103, 531–546.

Callen, J. (1991). Data envelopment analysis: practical survey and managerial accounting applications, *Journal of Management Accounting Research*, 3(1991), 35–57.

Charnes, A., Cooper, W. &. Rhodes, E. (1978). Measuring the efficiency of decision making units. *European Journal of Operations Research*, 2, 429–444.

Chen, H. & Kocaoglu, D. (2008). A sensitivity analysis algorithm for hierarchical decision models, *European Journal of Operations Research*, 185(1), 266–288.

Cooper, W., Li, S., Seiford, L., Thrall, R.M. & Zhu, J. (2001). Sensitivity and stability analysis in DEA: some recent developments, *Journal of Productivity Analysis*, 15(3), 217–246.

Cooper, W., Seiford, L. & Tone, K. (2000). *Data envelopment analysis*. Kluwer Academic Press, London, UK.

Fishburn, P.C. (1967). Additive utilities with incomplete product set: Applications to priorities and assignments. *Operations Research Society of America (ORSA)*, 15, 537–542.

Fox, W. (2018). *Mathematical modeling for business analytics*. Tayler and Francis Publishers, Boca Raton, FL.

Fox, W. & Burks, R. (2021). *Applied advanced mathematical modeling with technology*. Chapman and Hall/CRC, Boca Raton, FL.

Fox, W. P. (2020). Using multi-attribute decision making to rank the all-time greatest baseball player. *Baseball Research Journal*, 49(1), 106–113.

Giordano, F.R., Fox, W. & Horton, S. (2014). *A first course in mathematical modeling* (5th ed.). Brooks-Cole Publishers, Boston.

Hwang, C.L. & Yoon, K. (1981). *Multiple attribute decision making: Methods and applications: A state-of-the-art survey.* Springer-Verlag, Berlin and New York.

Krackhardt, D. (1990). Assessing the political landscape: structure, cognition, and power in organizations. *Administrative Science Quarterly,* 35, 342–369.

Leonelli, R. (2012). *Enhancing a decision support tool with sensitivity analysis.* Thesis, University of Manchester.

Saaty, T. (1980). *The analytic hierarchy process.* McGraw-Hill Book Company, New York, NY.

Saaty, T.L. (1990). How to make a decision: The analytic hierarchy process. *European Journal of Operational Research,* 48(48), 9–26.

Thanassoulis, E. (2011). *Introduction to the theory and application of data envelopment analysis-a foundation text with integrated software.* Kluwer Academic Press, London, UK.

Trick, M.A. (1996). *Multiple criteria decision making for consultants.* http://mat.gsia.cmu.edu/classes/mstc/multiple/multiple.html Accessed April 2014.

Trick, M.A. (2014). *Data envelopment analysis, chapter 12.* http://mat.gsia.cmu.edu/classes/QUANT/NOTES/chap12.pdf Accessed April 2014.

Winston, W. (1995). *Introduction to mathematical programming.* Duxbury Press, Belmont, CA, 322–325.

CHAPTER 11

Regression Techniques

R EGRESSION IS A STATISTICAL analysis technique for identifying the relationship between dependent and independent factors/variables. It helps decision-makers understand how the changes in independent factors might impact changes in the dependent factor of interest. Typically, practitioners use regression analysis to understand trends, determine the strength of predictors, and forecast potential impacts. Regression analysis, coupled with business analysis techniques, helps decision-makers analyze their data to support better-informed decisions. It is one of the most widely used data analysis approaches.

11.1 INTRODUCTION TO REGRESSION TECHNIQUES

Let's start the introduction to regression analysis with a Global Warming problem. Specifically, we will look at the level of carbon dioxide concentrations in the atmosphere and its impact on Global Warming. According to several news reports, the level of carbon dioxide in the atmosphere reached 379 parts per million (ppm) in March of 2004. It was stated that this was a larger increase than that observed during the past decade. The data provided in Table 11.1 is from a monitoring station at the Mauna Loa Observatory on the island of Hawaii. The station, located at an elevation of approximately 2 miles, is operated by the National Oceanic and Atmospheric Administration's climate-monitoring laboratory in Boulder, Colorado, United States.

It is estimated that the level of carbon dioxide in the atmosphere was approximately 280 parts per million before the industrial age began. There are some predictions that there will be a carbon dioxide

DOI: 10.1201/9781003464969-11

TABLE 11.1   Atmospheric CO_2 Concentrations

Year (CO_2 Reading for March of Year)	Parts per Million
1958	315.71
1959	316.65
1960	317.58
1961	318.54
1962	319.69
1963	319.86
1964	320.00
1965	320.89
1966	322.39
1967	323.04
1968	323.89
1969	325.64
1970	326.93
1971	327.18
1972	327.75
1973	330.30
1974	331.48
1975	332.04
1976	333.50
1977	334.70
1978	336.64
1979	337.96
1980	340.08
1981	341.38
1982	342.70
1983	343.10
1984	345.28
1985	347.43
1986	347.86
1987	349.42
1988	352.22
1989	353.68
1990	355.39
1991	357.16
1992	357.81
1993	358.38
1994	359.97
1995	361.64
1996	364.03

(*Continued*)

TABLE 11.1 *(Continued)*   Atmospheric CO_2 Concentrations

Year (CO_2 Reading for March of Year)	Parts per Million
1997	364.57
1998	367.31
1999	369.59
2000	370.52
2001	372.12
2002	373.52
2003	376.11
2004	377.70
2005	379.98
2006	382.09
2007	384.02
2008	385.83
2009	387.64
2010	390.10
2011	391.85
2012	394.06
2013	394.74
2014	398.81
2015	401.01
2016	404.41
2017	406.76
2018	408.72
2019	411.66
2020	414.24
2021	416.45

concentration in the atmosphere of 650 to 970 ppm by the year 2100 if growth continues at the current rate. We need to ask if this is a realistic predication. We will want to fit various mathematical models to the data and use the models to predict the CO_2 concentration in the atmosphere in the year 2100 and determine when the amount of carbon dioxide reaches 650 and 970 ppm.

Atmospheric CO_2 concentrations (ppmv) derived from in-situ air samples collected at Mauna Loa Observatory, Hawaii: http://cdiac.esd.ornl. gov/ftp/maunaloa-co$_2$/maunaloa.co$_2$

We desire to build a linear regression model to check the claims being made and will return to this example later in this chapter. First, let's briefly discuss regression.

We have data that relates to time to demand for a specific item for the past two years. We would like to forecast next year's demand, assuming things remain almost unchanged. Can we do this? If so, how do we proceed?

11.1.1 Why Regression Techniques?

Often, we have data, and we want or need to find a relationship between the independent and dependent variables. Notice that I have said relationship. Relationships are linear or nonlinear. The concept of correlation ONLY applies to linear relationships! Why do we obtain a model via regression? We might want to explain a behavior or predict a future or intermediate result. Several issues need to be examined besides just building or obtaining the model. These include: How good is the model we have built, do the results pass the commonsense test, and did I use the appropriate regression technique? These lessons help answer these important questions.

11.1.2 Correlation, Covariance, and Its Misconceptions

More people fail to understand the concept of correlation than anything else we do in regard to regression analysis. When two variables X and Y are not independent, it is often of interest to measure how strongly they are related to one another. Covariance, in many statistics books, has several formulas that we might use, such as

$$COV(X, Y) = E [XY] - \mu_x\mu_y \text{- or}$$
$$\Sigma x \Sigma y (x - \mu_x) (y - \mu_y) \times p(x,y)$$

Both use probability information.

11.1.3 Correlation: A Measure of Linear Relationship

Excel's definition & description is NOT correct as are many other definitions on the internet.

Returns the correlation coefficient of the array1 and array2 cell ranges. Use the correlation coefficient to determine the relationship between two properties. For example, you can examine the relationship between a location's average temperature and the use of air conditioners.

11.1.3.1 Correlation is the Measure of Linear Relationship between Variables

Correlation is computed into what is known as the correlation coefficient, which ranges between −1 and +1. Perfect positive correlation (a correlation coefficient of +1) implies that as one variable moves, either up or down, the other variable will move in lockstep in the same direction. Alternatively, perfect negative correlation means that if one variable moves in either direction, the variable that is perfectly negatively correlated will move in the opposite direction. If the correlation is 0, the movements of the variable are said to have no correlation; they are completely random, or their relationship is nonlinear.

In real life, perfectly correlated variables are rare; rather you will find variables with some degree of correlation.

In the textbook by *Devore, Probability and Statistics for Engineering and Sciences*, he says that for descriptive purpose, the correlation relationship, called ρ, that measures the linear relationship of the variables will be described as:

Strong when $|\rho| > 0.8$,

Moderate if $0.5 < |\rho| < 0.8$,

and Weak if $|\rho| < 0.5$.

For social science data, the correlation measures scales are different (Table 11.2).

Caution must be used when discussing correlation. We provide two examples (Figures 11.1 and 11.2) both that show that there is perhaps a relationship between x and y but not always it is linear.

The correlation between x and y in Figure 11.1 is 0.779. The plot appears reasonably linear.

The correlation between x and y in Figure 11.2 is 0.108. The plot is not remotely linear; yet, clearly, we see the oscillating relationship between x and y.

TABLE 11.2 Correlation Level Strengths

Correlation Level	Values
None	$0.0 \leq |\rho| \leq 0.1$
Small	$0.1 < |\rho| \leq 0.3$
Medium	$0.3 < |\rho| \leq 0.5$
Strong	$0.5 < |\rho| \leq 1$

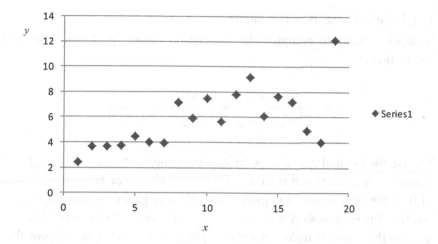

FIGURE 11.1   Strong Correlation and Linear Relationship.

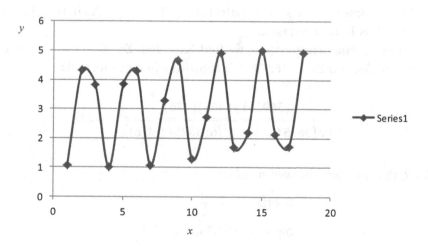

FIGURE 11.2   Weak Correlation But Clear Relationship.

11.1.3.2 Correlation – A Measure of Linear Relationship

Correlation is one of the most common and most useful statistics. A correlation is a single number that describes the degree of linear relationship between two variables. Let's work through an example to show you how this statistic is computed.

11.1.4 Calculating the Correlation

Now we're ready to compute the correlation value. The formula for the correlation is:

$$\rho = \frac{n\sum xy - (\sum x)(\sum y)}{\sqrt{\left[n\sum x^2 - (\sum x)^2\right]\left[n\sum y^2 - (\sum y)^2\right]}}$$

We use the symbol r or ρ (rho) to stand for the correlation. Through the magic of mathematics, it turns out that ρ will always be between −1.0 and +1.0. If the correlation is negative, we have a negative relationship; if it's positive, the relationship is positive. You don't need to know how we came up with this formula unless you want to be a statistician. But you probably will need to know how the formula relates to real data – how you can use the formula to compute the correlation.

Let's assume we have six data pairs (Table 11.3) for grades in two classes. Let x = class 1 and y = class 2.

Through data manipulation, we find $\sum x = 464$, $\sum x^2 = 36354$, $\sum y = 476$, $\sum y^2 = 38,254$, and $\sum xy = 36,926$. We substitute into our formula for ρ.

$$\rho = \frac{6(36926) - (464)(476)}{\sqrt{[6(36354) - (464)^2(6(38254)(476)^2}} = 0.2396$$

In R, this is easier. The command is:

```
>cor(x,y)
mg = c(70,92,80,74,65,83)
> eg = c(74,84,63,87,78,90)
> cor(mg,eg)
[1] 0.2396639
```

TABLE 11.3 Data Pairs

x	70	92	80	74	65	83
y	74	84	63	87	78	90

If this data were scientific data, our correlation indicates a weak linear relationship, but if our data was social science data (nonscientific data), the correlation indicates a small linear relationship.

11.1.5 Example 11.1: Correlation for Global Warming Data

We use Excel and the command correl(array1,array2).

The correlation coefficient r is 0.991264. This indicates a strong linear relationship for this scientific data.

11.1.6 Testing the Significance of a Correlation with Hypothesis Testing

Once you've computed a correlation, you can determine the probability that the observed correlation occurred by chance. That is, you can conduct a significance test. Most often, you are interested in determining the probability that the correlation is a real one and not a chance occurrence. In this case, you are testing the mutually exclusive hypotheses:

Null Hypothesis: $\rho = 0$

Alternative Hypothesis: $\rho \neq 0$

The easiest way to test this hypothesis is to find a statistics book that has a table of critical values of ρ. Most introductory statistics texts would have a table like this, and since we are not using one, we include it here (Table 11.4).

TABLE 11.4   Critical Values for Pearson Correlation – N.

| | One-Tailed Probabilities | | | |
	0.05	0.025	0.005	0.0005
	Two-Tailed Probabilities			
N	0.100	0.050	0.010	0.001
4	0.900	0.950	0.990	0.999
5	0.805	0.878	0.959	0.991
6	0.729	0.811	0.917	0.974
7	0.669	0.754	0.875	0.951
8	0.621	0.707	0.834	0.925
9	0.582	0.666	0.798	0.898

TABLE 11.1 *(Continued)* Critical Values for Pearson Correlation – *N*.

	One-Tailed Probabilities			
10	0.549	0.632	0.765	0.872
11	0.521	0.602	0.735	0.847
12	0.497	0.576	0.708	0.823
13	0.476	0.553	0.684	0.801
14	0.458	0.532	0.661	0.780
15	0.441	0.514	0.641	0.760
16	0.426	0.497	0.623	0.742
17	0.412	0.482	0.606	0.725
18	0.400	0.468	0.590	0.708
19	0.389	0.456	0.575	0.693
20	0.378	0.444	0.561	0.679
21	0.369	0.433	0.549	0.665
22	0.360	0.423	0.537	0.652
23	0.352	0.413	0.526	0.640
24	0.344	0.404	0.515	0.629
25	0.337	0.396	0.505	0.618
26	0.330	0.388	0.496	0.607
27	0.323	0.381	0.487	0.597
28	0.317	0.374	0.479	0.588
29	0.311	0.367	0.471	0.579
30	0.306	0.361	0.463	0.570
35	0.283	0.334	0.430	0.532
40	0.264	0.312	0.403	0.501
45	0.248	0.294	0.380	0.474
50	0.235	0.279	0.361	0.451
60	0.214	0.254	0.330	0.414
70	0.198	0.235	0.306	0.385
80	0.185	0.220	0.286	0.361
90	0.174	0.207	0.270	0.341
100	0.165	0.197	0.256	0.324
200	0.117	0.139	0.182	0.231
300	0.095	0.113	0.149	0.189
400	0.082	0.098	0.129	0.164
500	0.074	0.088	0.115	0.147
1000	0.052	0.062	0.081	0.104

As in all hypothesis testing, you need to first determine the significance level. Here, we use the common significance level of alpha = 0.05. This means that we are conducting a test where the odds that the correlation is a chance occurrence is no more than 5 out of 100. Before we look up the critical values in Table 11.4, we also have to compute the degrees of freedom or *df*. The *df* is simply equal to $N - 2$ where N is the number of data pairs, or, in the Global Warming example, this is $df = 63 - 2 = 61$. Finally, we have to decide whether we are doing a one-tailed or two-tailed test. In this example, since we have no strong prior theory to suggest whether the relationship between height and self-esteem would be positive or negative, we will opt for the two-tailed test. With these three pieces of information – the significance level ($\alpha = 0.05$), degrees of freedom ($df = 61$), and type of test (two-tailed) – we can now test the significance of the correlation we found. When we look up this value in the table at the back of my statistics book, we find that the closest critical value is 0.213. This means that if the correlation is greater than 0.213 or less than −.213 (remember, this is a two-tailed test), we can conclude that the odds are less than 5 out of 100 that this is a chance occurrence. Since our correlation is 0.9913 is actually quite a bit higher, we conclude that it is not a chance finding and that the correlation is "statistically significant" (given the parameters of the test). We can reject the null hypothesis in favor of the alternative hypothesis.

11.2 MODEL FITTING AND LEAST SQUARES

Consider a simple spring-mass system where we examine the spring both at rest and with a weight attached to the end of it. We conducted an experiment to measure the stretch of the spring as a function of mass placed on the spring. We measure only how far the spring stretched from its original position, and this data is displayed in Table 11.5.

TABLE 11.5 Spring-Mass System

Mass (Grams)	Stretch (m)
50	0.100
100	0.188
150	0.275
200	0.325
250	0.438
300	0.488
350	0.568
400	0.650
450	0.725
500	0.800
550	0.875

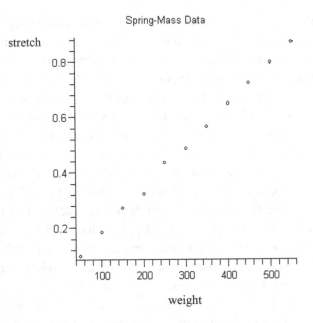

FIGURE 11.3   Plot of Spring-Mass data that Appears Linear.

The data plot, seen in Figure 11.3, looks reasonably like a straight line through the origin. Our next step was to calculate the slope. We use the points (50, 0.1) and (550, 0.875) and compute the slope as 0.00155. We find the model as $F = 0.00155$. We now want a more exact fit of our line to the data. Model fitting, especially with the least squares method, will be the way how we obtain a better fit.

This section focuses on the analytical methods to arrive at a model for a given dataset using a prescribed criterion. Again, from the family $y = kx^2$, the parameter k can be determined analytically by using a curve-fitting criterion, such as least squares, Chebyshev's, or minimizing the sum of the absolute error and then solving the resulting optimization problem. We concentrate on the presentation of least squares in this chapter. We present the R commands that solve the least-squares optimization problem with an analysis of the "goodness of the fit" of the resulting model.

Remark: In R we can obtain least squares fit of our data.

11.2.1 Example 11.2: Spring-Mass System in R

Mass = c(50,100,150,200,250,300,350,400,450,500,550)

Stretch = c(.1, .1875, .275, .325, .4375, .4875, .5675, .65,.725,.8,.875)

```
cor(Mass,Stretch)
[1] 0.9992718

> plot(Mass,Stretch)
```

Model by R:

```
> fit<-lm(Stretch~Mass)
> fit
```

Call:

lm(formula = Stretch ~ Mass)

Coefficients:
(Intercept) Mass

0.032455 0.00153

The model is Stretch = 0.02455 + 0.00153 Mass. We can use *R* to overlay the data and the least squares line (Figure 11.5).

We go ahead, at this time, and get the summary output from *R*.

```
> summary(fit)
```

Call:

lm(formula = Stretch ~ Mass)

Residuals:

Min 1Q Median 3Q Max

− 0.014909−0.004568−0.001091 0.001977 0.020727

Coefficients:

Estimate Std. Error t value Pr(>|t|)

(Intercept) 3.245e-02 6.635e-03 4.891 0.000858 ***

Mass 1.537e-03 1.957e-05 78.569 4.44e-14 ***

− --

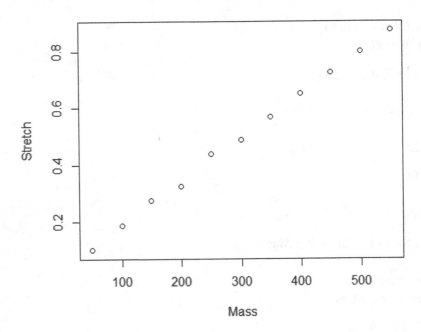

FIGURE 11.4   Plot of Data for Example 11.2.

Signif. codes: 0 '***' 0.001 '**' 0.01 '*' 0.05 '.' 0.1 ' ' 1

Residual standard error: 0.01026 on 9 degrees of freedom

Multiple R-squared: 0.9985,   Adjusted R-squared: 0.9984

F-statistic: 6173 on 1 and 9 df, p-value: 4.437e-14

11.2.2 Global Warming Example 11.1

We can now return to our Global Warming problem that started this chapter. The plot (Figure 11.6) of data appears linear to support the correlation found earlier.

We used the data analysis add-in in Excel. The following two screenshots (Figures 11.7 and 11.8) provide the results.

We obtained the residual list screenshot in Figure 11.7.

The residual plot is shown in Figure 11.9.

We see a curved trend, so the model is not adequate. We should try other regression approaches to improve the model.

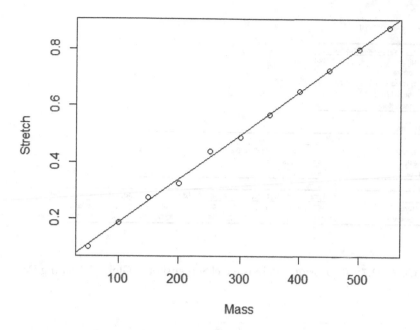

FIGURE 11.5   Least Square Line Fit to Data for Example 11.2.

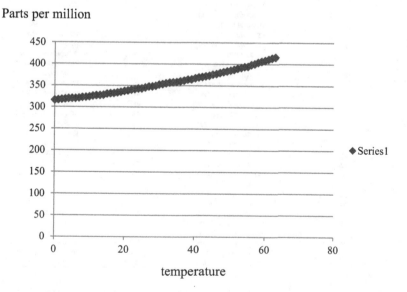

FIGURE 11.6   Plot of Global Warming Data.

SUMMARY OUTPUT

Regression Statistics	
Multiple R	0.991264
R Square	0.982604
Adjusted R Square	0.982323
Standard Error	3.948497
Observations	64

ANOVA

	df	SS	MS	F	Significance F
Regression	1	54598.06	54598.06	3501.98	2.89623E-56
Residual	62	966.6188	15.59063		
Total	63	55564.68			

	Coefficient	Standard Err	t Stat	P-value	Lower 95%	Upper 95%	Lower 95.0%	Upper 95.0%
Intercept	307.4403	0.975668	315.1076	4.5E-101	305.48999	309.3907	305.49	309.3907
Year scaled	1.581111	0.026718	59.17753	2.9E-56	1.527702098	1.63452	1.527702	1.63452

FIGURE 11.7   Regression Analysis Excel Screenshot of Global Warming Data.

Observation	Predicted Parts per Million	Residuals	obser.	Predicted Parts per Million	Residual
1	307.4403221	8.269678	33	358.0358679	-2.64587
2	309.0214329	7.628567	34	359.6169787	-2.45698
3	310.6025437	6.977456	35	361.1980895	-3.38809
4	312.1836545	6.356345	36	362.7792003	-4.3992
5	313.7647653	5.925235	37	364.3603111	-4.39031
6	315.3458761	4.514124	38	365.9414219	-4.30142
7	316.926987	3.073013	39	367.5225327	-3.49253
8	318.5080978	2.381902	40	369.1036435	-4.53364
9	320.0892086	2.300791	41	370.6847543	-3.37475
10	321.6703194	1.369681	42	372.2658652	-2.67587
11	323.2514302	0.63857	43	373.846976	-3.32698
12	324.832541	0.807459	44	375.4280868	-3.30809
13	326.4136518	0.516348	45	377.0091976	-3.4892
14	327.9947626	-0.81476	46	378.5903084	-2.48031
15	329.5758734	-1.82587	47	380.1714192	-2.47142
16	331.1569842	-0.85698	48	381.75253	-1.77253
17	332.738095	-1.2581	49	383.3336408	-1.24364
18	334.3192058	-2.27921	50	384.9147516	-0.89475
19	335.9003166	-2.40032	51	386.4958624	-0.66586
20	337.4814274	-2.78143	52	388.0769732	-0.43697
21	339.0625382	-2.42254	53	389.658084	0.441916
22	340.643649	-2.68365	54	391.2391948	0.610805
23	342.2247598	-2.14476	55	392.8203056	1.239694
24	343.8058707	-2.42587	56	394.4014164	0.338584
25	345.3869815	-2.68698	57	395.9825272	2.827473
26	346.9680923	-3.86809	58	397.563638	3.446362
27	348.5492031	-3.2692	59	399.1447489	5.265251
28	350.1303139	-2.70031	60	400.7258597	6.03414
29	351.7114247	-3.85142	61	402.3069705	6.41303
30	353.2925355	-3.87254	62	403.8880813	7.771919
31	354.8736463	-2.65365	63	405.4691921	8.770808
32	356.4547571	-2.77476	64	407.0503029	9.399697

FIGURE 11.8   Regression Analysis Residuals of Global Warming Data.

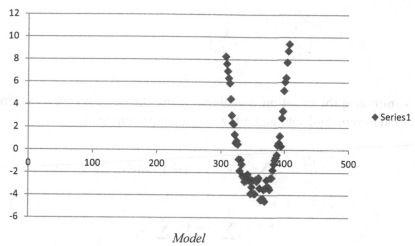

FIGURE 11.9   Plot of Regression Analysis Residuals for Global Warming Data.

11.3 THE DIFFERENT CURVE-FITTING CRITERION

We will briefly cover our curve-fitting criterion for least squares or linear regression.

11.3.1 Criterion 1: Least Squares

The method of least-squares curve fitting, also known as *ordinary least squares* and *linear regression*, is simply the solution to a model that minimizes the sum of the squares of the deviations between the observations and predictions. Least squares will find the parameters of the function $f(x)$ that will

minimize $$S = \sum_{j=1}^{m} \left[y_1 - f(x_j) \right]^2$$

For example, to fit a proposed model $y = kx^2$ to a set of data, the least-squares criterion requires the minimization of the function, where k is a slope.

Minimize

$$S = \sum_{j=1}^{5} \left[y_i - kx_j^2 \right]^2$$

Minimizing the equation is achieved using the first derivative, setting it equal to zero, and solving for the unknown parameter, k.

$$\frac{ds}{dk} = -2 \sum x_j^2 (y_j - kx_j^2) = 0. \text{ Solving for k:}$$

$$k = \left(\sum x_j^2 y_j \right) / \left(\sum x_j^4 \right).$$

Given the dataset in Table 11.6, we will find the least squares fit to the model, $y = kx^2$.

Solving for k: $k = \left(\sum x_j^2 y_j \right) / \left(\sum x_j^4 \right) = (195.0) / (61.1875) = 3.1869$ and the model $y = kx^2$ becomes $y = 3.1869x^2$

Let's assume we prefer the model, $y = ax2 + bx + c$

$$xv := [.5, 1, 1.5, 2, 2.5];$$

$$yv := [.7, 3.4, 7.2, 12.4, 20.1];$$

The least-squares fit is:

$$y = 3.2607x^2 - 0.22223x + 0.12630$$

Next, we illustrate the least squares fit applied to our model $y = kx^2$ for the dataset for our example. As obtained previously, the least-squares

TABLE 11.6 Least-Squares Data Points

X	0.5	1	1.5	2	2.5
Y	0.7	3.4	7.2	12.4	20.1

model is $y = 3.1870x^2$ (rounded to 3 decimals places). The fit will not be as good nor appear as good although this might not always be visible to the eye. Why?

11.3.2 Example 11.3: Least-Squares Fit Explosive Data

Assume that we have developed a proportionality model, $V = kD^3$. Let's demonstrate how to determine a constant of proportionality, using the least squares criterion. V represents the volume of the crater from TNT and diameter³ is the estimate of the volume of the crater created. This example illustrates the fit command, analytically fitting the model $V = k$ diameter³ to the same dataset using our data in Table 11.7.

Our least squares model is:

$$y = 0.0084011\ x^3.$$

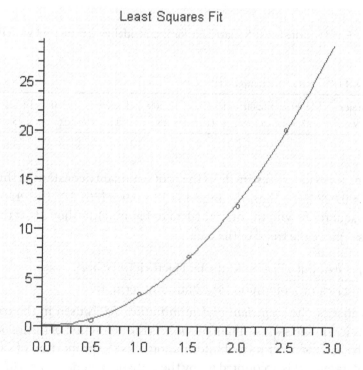

FIGURE 11.10   Least-Squares Fit Plotted with the Original Data.

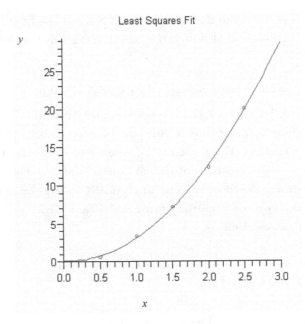

FIGURE 11.11 The Least-Squares Fit for the Model $y = kx^2$ plotted with the data.

TABLE 11.7 Data for Example 11.2

Diameter	14.500	12.500	17.250	14.500	12.625	17.750	14.125	12.625
Size, V	27	17	41	26	17	49	23	16

The least-squares estimate of the proportionality constant in this model is $k = 0.0084365$. Thus, the model is $V = 0.0084365\ D^3$. The graph of the least squares fit with the original data in Figure 11.12 shows that the model does capture the trend of the data.

11.4 DIAGNOSTICS AND INTERPRETATIONS

Coefficient of Determination: Statistical term: R^2

In statistics, the *coefficient of determination, R^2*, is used in the context of statistical models whose main purpose is the prediction of future outcomes on the basis of other related information. It is the proportion of variability in a dataset that is accounted for by the statistical model. It provides a measure of how well future outcomes are likely to be predicted by the model.

There are several different definitions of R^2 which are only sometimes equivalent. One class of such cases includes that of linear regression.

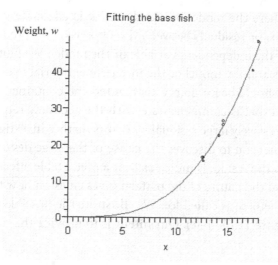

FIGURE 11.12   Fitting Least Square Model to Example 11.2 Data.

In this case, R^2 is simply the square of the sample correlation coefficient between the outcomes and their predicted values, or in the case of simple linear regression, between the outcome and the values being used for prediction. In such cases, the values vary from 0 to 1. Important cases where the computational definition of R^2 can yield negative values, depending on the definition used, arise where the predictions which are being compared to the corresponding outcome have not been derived from a model-fitting procedure using those data.

$$R^2 = 1 - SSE/SST$$

Values of R^2 outside the range 0 to 1 can occur where they are used to measure the agreement between observed and modeled values and where the "modeled" values are not obtained by linear regression and depending on which the formulation of R^2 is used. If the first formula above is used, values can never be greater than 1.

11.4.1 Plotting the Residuals for a Least-Squares Fit

In the previous section, you learned how to obtain a least squares fit of a model, and you plotted the model's predictions on the same graph as the observed data points in order to get a visual indication of how well the model matches the trend of the data. A powerful technique for quickly determining where the model breaks down or is adequate is to plot the actual deviations or residuals between the observed and predicted values as a function of the independent variable or the model. We plot the residuals in the y-axis and the model or the independent variable on the x-axis. The deviations should be randomly distributed and contained in a reasonably small band that is commensurate with the accuracy required by the model. Any excessively large residual warrants further investigation of the data point in question to discover the cause of the large deviation. A pattern or trend in the residuals indicates that a predictable effect remains to be modeled, and the nature of the pattern gives clues on how to refine the model, if a refinement is called for. We illustrate the possible patterns for residuals in Figure 11.13(a)–(e). Our intent is to provide the modeler with

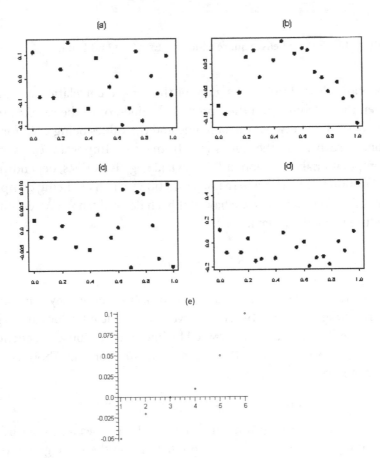

FIGURE 11.13 Patterns for Residuals: (a) No Pattern, (b) Curved Pattern, (c) Fanning Pattern,

(D) OUTLIERS, AND (E) LINEAR TREND.

knowledge concerning the adequacy of the model they have found. We will leave further investigations into correcting the patterns to follow on courses in statistical regression.

In Excel, we can always request a residual plot. After fitting a specified model, the difference between the observed and predicted values can be calculated by using array manipulations.

11.4.2 Example 11.4

We return to our explosive example (Example 11.2) from the Section 11.3.1. In this example, the relationship was modeled by the following expression:

$$V = 0.00854365 \ diameter^3$$

We use our data from Table 11.8 to build a regression model.

The residuals are the differences between the predicted and observed values. Mathematically, these are called errors, deviations, or residuals and are found by

$$y_i - f(x_i)$$

In our example, the errors are found using the model

$$V_i - 0.00854365 * D_i^3$$

and are 1.388, 0.592, –2.498, 0.388, 0.194, 1.619, –0.550, and –0.806. We plot these errors versus the model (Figure 11.14).

Note that the residuals are randomly distributed and contained in a relatively small band about zero. There are no outliers, or unusually large residuals, and there appears to be no pattern in the residuals. Based on these aspects of the plot of the residuals, the model approximates the data.

11.4.3 Example 11.5: Fruit Flies over Time

For this next example, we will fit a cubic equation to the following data in Table 11.9.

TABLE 11.8   Data for Example 11.3

Diameter	14.500	12.500	17.250	14.500	12.625	17.750	14.125	12.625
Size, V	27	17	41	26	17	49	23	16

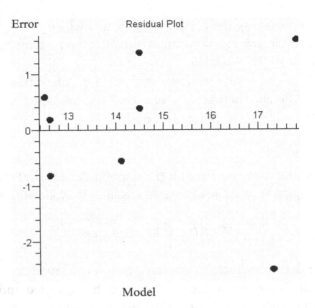

FIGURE 11.14   The Explosive Model's Residuals.

TABLE 11.9   Data for Example 11.4

T	7	14	21	28	35	42
P	125	275	800	1,200	1,700	1,650

There appears to be a change in concavity in the data, so we decided to try a cubic equation. The least squares model is approximately:

$$Y = -0.1066x^3 + 7.436x^2 - 95.7814x + 466.6667$$

We use this model, and we find the residuals (Figure 11.16) as *0.9920635, −15.6746, 52.7777, −74.206, 47.817, −11.706.*

There does not appear to be any pattern to the residual plot (Figure 11.16), so we conclude the model is adequate.

11.4.3.1 Percent Relative Error

When using a model to predict information, we really want to know how well the model appears to work. We will use percent relative error (% REL ERR).

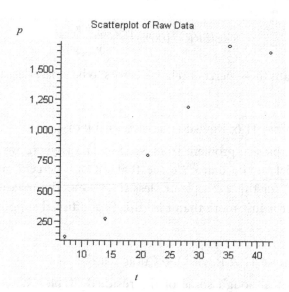

FIGURE 11.15 Plot of Raw Data in Table 11.9 for Example 11.4

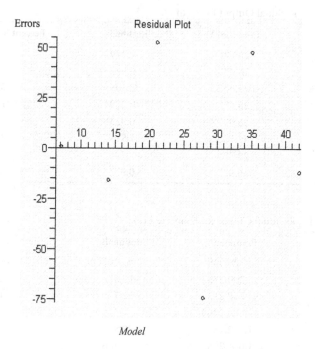

Model

FIGURE 11.16 Residual Plot for Least Square Model for Example 2.

$$\% \operatorname{Re}lERR = 100\% \cdot \frac{\mid y_{actual} - y_{model} \mid}{y_{actual}}$$

We really want these percent relative errors to be small (less than 10–20% on average).

11.4.4 Example 11.6: Revisit Explosive Problem

Recall the explosive problem from Section 11.3.1 where we fitted a least square model to the data. We see that all our percent relative errors (Table 11.10) for this example are less than 10%. The maximum percent relative error is just more than 6%. This is additional support for a good model.

11.4.5 Example 11.7: Revisit the Cubic Model

We note that although some of the residual (Table 11.11) appear large (52.77 and 74.206), all the percent relative errors are less than 7%, and support for a good model.

TABLE 11.10   Residual Output for Example 11.5

Observation	Predicted Y	Residuals	Percent Relative Errors
1	25.721	1.279	4.739
2	16.478	0.522	3.070
3	43.305	−2.305	5.623
4	25.721	0.279	1.075
5	16.977	0.023	0.133
6	47.181	1.819	3.711
7	23.776	−0.776	3.374
8	16.977	−0.977	6.108

TABLE 11.11   Residual Output for Example 11.6

Observation	Predicted Y	Residuals	Percent Relative Error
1	124.008	0.992	0.794
2	290.675	−15.675	5.700
3	747.222	52.778	6.597
4	1,274.206	−74.206	6.184
5	1,652.183	47.817	2.813
6	1,661.706	−11.706	0.709

11.5 DIAGNOSTICS AND INFERENTIAL STATISTICS

While descriptive statistics is used by OR practitioners to summarize the characteristics of a set of data, inferential statistics help the OR practitioner to make conclusions and potential predictions based on changes and relationship in the data. Inferential statistics will allow you to understand the larger population based on the smaller set of collected sample data. This section will explore the principles of inferential statistics using R.

11.5.1 The Spring-Mass System Using R
R Code:

> summary(fit)

Call:

lm(formula = Stretch ~ Mass)

Residuals:

Min 1Q Median 3Q Max

– 0.014909–0.004568–0.001091 0.001977 0.020727

Coefficients:

Estimate Std. Error t value Pr(>|t|)

(Intercept) 3.245e-02 6.635e-03 4.891 0.000858 ***

Mass 1.537e-03 1.957e-05 78.569 4.44e-14 ***

– --

Signif. codes: 0 '***' 0.001 '**' 0.01 '*' 0.05 '.' 0.1 ' ' 1

Residual standard error: 0.01026 on 9 degrees of freedom
Testing the coefficients of the model with hypothesis testing.
The hypothesis tests being run are:

Ho: $B_i = 0$

Ha: $B_i \neq 0$

for the slope and the intercept. The p-values for the slope (4.44e-14) and the intercept (0.000858) are both less than 0.05 (our alpha level), and we conclude that both are nonzero and need to be part of the equation.

11.5.2 Simple Linear Regression Model with Complete Explanation Summary in R

We are interested in trying to develop a model to make a prediction about the amount of wheat (in Bushels) that we are growing on our farm. Table 11.12 provides our production for the last 20 years.

We can use R to conduct our regression analysis.

R Code:

```
>year=c(1,2,3,4,5,6,7,8,9,10,11,121,31,41,5,16,17,18,19,20)
> quantity=c(50,47,51,46,45,44,46,38,39,37,36,32,30,32,30,28,26,24,
25,22)
> fit_1<-lm(quantity~year)
> fit_1
```

Call:

```
lm(formula = quantity ~ year)
```

Coefficients:

```
(Intercept) year
52.537 -1.537
> summary(fit_1)
```

Call:

```
lm(formula = quantity ~ year)
```

TABLE 11.12   Wheat Yield per Year

Year	1	2	3	4	5	6	7	8	9	10	11	12	13	14	15	16	17	18	19	20
YYield (Bushels)	50	47	51	46	45	44	46	38	39	37	36	32	30	32	30	28	26	24	25	22

Residuals:

Min 1Q Median 3Q Max

− 2.5579−0.9053 0.1000 0.5579 4.2211

Coefficients:

Estimate Std. Error t value Pr(>|t|)

(Intercept) 52.53684 0.82043 64.03 < 2e-16 ***

year -1.53684 0.06849 -22.44 1.31e-14 ***

− --

Signif. codes: 0 '***' 0.001 '**' 0.01 '*' 0.05 '.' 0.1 ' ' 1

Residual standard error: 1.766 on 18 degrees of freedom

Multiple R-squared: 0.9655, Adjusted R-squared: 0.9636

F-statistic: 503.5 on 1 and 18 df, p-value: 1.309e-14

> plot(year,quantity)

> abline(fit_1)

The model is written as *quantity = 52.53684 − 1.5684 year*
We see in Figure 11.17 that this linear model appears to fit the data quite well.
Diagnostics: Residual plot (Figure 11.18) – appears pretty random. Hypothesis tests results all show significance, which is good.
Residuals:

Min 1Q	Median	3Q	Max
− 2.5579	−0.9053	0.1000	0.5579 4.2211

Coefficients:

Estimate Std. Error t value Pr(>|t|)

(Intercept) 5 2.53684 0.82043 64.03 < 2e-16 ***

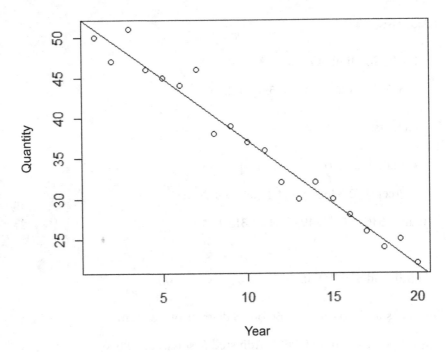

FIGURE 11.17   Line and Data Overlaid.

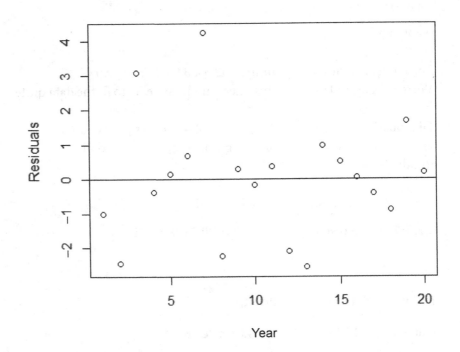

FIGURE 11.18   Residual Plot.

year −1.53684 0.06849 −22.44 1.31e-14 ***

– --

Signif. codes: 0 '***'

Residual standard error: 1.766 on 18 degrees of freedom

Multiple *R*-squared: 0.9655,   Adjusted *R*-squared: 0.9636

F-statistic: 503.5 on 1 and 18 *df*, p-value: 1.309e-14 ← Model P-value is
significant.
SSE and SSR

sse = sum((fitted(fit_1) − mean(quantity))^2)

> ssr = sum((fitted(fit_1) − quantity)^2)

> 1 − (ssr/(sse + ssr))

[1] 0.965486

>

> sse

[1] 1570.653

> ssr

[1] 56.14737

11.6 POLYNOMIAL REGRESSION IN R

Polynomial regression is one form of regression analysis in which the relationship between the dependent factor and independent factors is modeled with a polynomial. Polynomial regression is a special form of linear regression where due to the nonlinear relationship between dependent and independent factors, we add a polynomial term to the linear regression to convert it into polynomial regression.

11.6.1 Example 11.8: Recovery Level versus Time

We have the following data for time and levels of results. The plot of the data is in Figure 11.19, and you can see a curved trend.

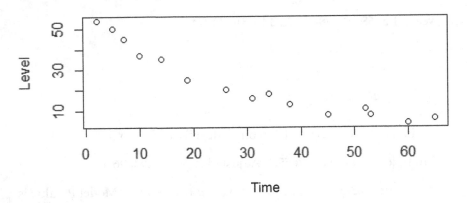

FIGURE 11.19   Scatterplot of Data, Time versus Level.

R Code:

```
> Level=c(54,50,45,37,35,25,20,16,18,13,8,11,8,4,6)
> time=c(2,5,7,10,14,19,26,31,34,38,45,52,53,60,65)
> plot(time, Level)
```

We obtain a correlation value in R.

R Code

```
cor(time, Level)
[1] -0.9410528
```

This implies a strong linear relationship. However, due to the curve trend we see in the scatterplot, we decided to model with a quadratics function.
recovery_model2 < – lm(Level ~ time + I((time)^2 +) > summary (recovery_model2)
Call:

lm(formula = Level ~ time + I((time)^2))

Residuals:

Min 1Q Median 3Q Max

– 3.6724 –1.2896 0.2194 1.3841 4.0736

Coefficients:

Estimate Std. Error t value Pr(>|t|)

(Intercept) 55.822134 1.649202 33.848 2.81e-13 ***

time −1.710263 0.124797−13.704 1.09e-08 ***

I((time)^2) 0.014807 0.001868 7.927 4.13e-06 ***

– --

Signif. codes: 0 '***' 0.001 '**' 0.01 '*' 0.05 '.' 0.1 ' ' 1

Residual standard error: 2.455 on 12 degrees of freedom

Multiple R-squared: 0.9817,　　Adjusted R-squared: 0.9786

F-statistic: 321.1 on 2 and 12 df, p-value: 3.812e-11

The model is recovery level = 55.822123−1.700263 time + 0.014807 time^2.

Residual plot is shown in Figure 11.20.

sse

[1] 3870.991

> ssr

[1] 56.14737

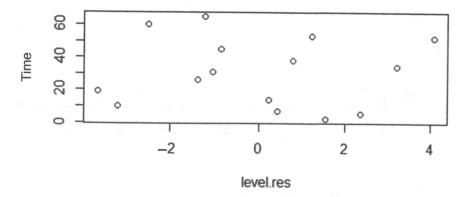

FIGURE 11.20　Residual Plot.

> 1 – (ssr/(sse + ssr))

[1] 0.9857027

Percent relative error:

Per_rel_err

| 1 | 2 | 3 | 4 | 5 | 6 |

– 2.8503045 –4.7180144 –0.9425917 8.6491871 –0.6268164 14.6897820

| 7 | 8 | 9 | 10 | 11 | 12 |

6.8239618 6.4591042 –17.8332119 –6.0510012 10.5545422 –37.0323886

| 13 | 14 | 15 |

–15.3637082 62.7838691 20.2396576

From all indicators, it appears to be a good model. However, when we go to predict the levels after 100 days, we obtain an answer that does not pass a common-sense test. We expect the value of levels to get closer to 0; however, $f(100) = 33.82$.

```
fm <- lm(Level ~ poly(time, 2), BOD)
> plot(Level ~time, BOD)
> lines(fitted(fm) ~ time, BOD, col = "red")
```

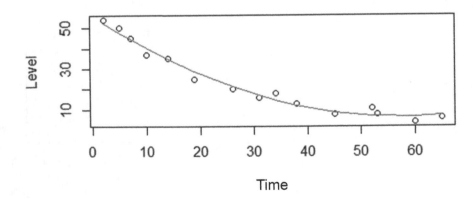

FIGURE 11.21   Fitted Line to Data.

11.6.2 Wheat Production Revisited

Let's return to our wheat production problem from Section 11.5.2. The data was provided in Table 11.7, but instead of modeling it as quantity = function (time) as a linear model, we desire a quadratic polynomial such as quantity = b_0 + $b_1$time + $b_2$year^2.

We will use all the same diagnostics as before.

R Code

Model = lm(quantity~poly(year,2))

> summary(model)

Call:

lm(formula = quantity ~ poly(year, 2))

Residuals:

Min 1Q Median 3Q Max

− 2.5271−0.9411 0.0956 0.5727 4.2450

Coefficients:

Estimate Std. Error t value Pr(>|t|)

(Intercept) 36.4000 0.4063 89.591 < 2e-16 ***

poly(year, 2)1−39.6315 1.8170−21.812 7.22e-14 ***

poly(year, 2)2 0.1509 1.8170 0.083 0.935

− --

Signif. codes: 0 '***' 0.001 '**' 0.01 '*' 0.05 '.' 0.1 ' ' 1

Residual standard error: 1.817 on 17 degrees of freedom

Multiple R-squared: 0.9655,   Adjusted R-squared: 0.9614

F-statistic: 237.9 on 2 and 17 df, p-value: 3.728e-13

11.7 EXERCISES

11.1. Find the correlation to fit the data in Table 11.13 with the models using least squares:

TABLE 11.13   Data for Exercise 11.1

X	1	2	3	4	5
Y	1	1	2	2	4

    a. $y = b + ax$

    b. $y = ax^2$

11.2. Find the correlation of the data for the stretch of spring data in Table 11.14 and the fit with the models using least squares.

TABLE 11.14   Spring Stretch Data for Exercise 11.2

$x(x\,10^{-3})$	5	10	20	30	40	50	60	70	80	90	100
$x(x\,10^5)$	0	19	57	94	134	173	216	256	297	343	390

    a. $y = ax$

    b. $y = b + ax$

    c. $y = ax^2$

11.3. Data for the ponderosa pine are given in Table 11.15.

TABLE 11.15   Poderosa Pine Data for Exercise 11.3

X	17	19	20	22	23	25	28	31	32	33	36	37	39	42
y	19	25	32	51	57	71	113	140	153	187	192	205	250	260

    a. $y = ax + b$

    b. $y = ax^2$

    c. $y = ax^3$

    d. $y = ax^3 + bx^2 + c$

11.4. Scientists think that there is a relationship between Global Warming temperatures and the amount of carbon dioxide. Find the correlation and least squares fit.

Over the same period, the land ocean temperature index measured in Celsius is as follows (modified from https://data.giss.nasa.gov/gistemp/graphs/graph_data/Global_Mean_Estimates_based_on_Land_and_Ocean_Data/graph.txt):

1951 −0.07 −0.07	1975 −0.01 0.02	1999 0.38 0.47
1952 0.01 −0.07	1976 −0.10 0.04	2000 0.39 0.50
1953 0.08 −0.07	1977 0.18 0.07	2001 0.53 0.52
1954 −0.13 −0.07	1978 0.07 0.12	2002 0.62 0.54
1955 −0.14 −0.06	1979 0.16 0.16	2003 0.61 0.58
1956 −0.19 −0.05	1980 0.26 0.20	2004 0.53 0.60
1957 0.05 −0.04	1981 0.32 0.21	2005 0.67 0.61
1958 0.06 −0.01	1982 0.14 0.22	2006 0.63 0.62
1959 0.03 0.01	1983 0.31 0.21	2007 0.66 0.63
1960 −0.03 0.03	1984 0.16 0.21	2008 0.54 0.64
1961 0.06 0.01	1985 0.12 0.22	2009 0.65 0.64
1962 0.03 −0.01	1986 0.18 0.24	2010 0.72 0.64
1963 0.05 −0.03	1987 0.32 0.27	2011 0.61 0.66
1964 −0.20 −0.04	1988 0.39 0.31	2012 0.64 0.69
1965 −0.11 −0.05	1989 0.27 0.33	2013 0.67 0.73
1966 −0.06 −0.06	1990 0.45 0.33	2014 0.74 0.78
1967 −0.02 −0.05	1991 0.40 0.33	2015 0.90 0.83
1968 −0.08 −0.03	1992 0.22 0.33	2016 1.01 0.87
1969 0.05 −0.02	1993 0.23 0.33	2017 0.92 0.91
1970 0.02 −0.01	1994 0.31 0.34	2018 0.85 0.92
1971 −0.08 0.00	1995 0.45 0.37	2019 0.97 0.92
1972 0.01 0.00	1996 0.33 0.40	2020 1.01 0.93
1973 0.16 −0.00	1997 0.46 0.42	2021 0.84 0.93
1974 −0.07 0.01	1998 0.61 0.44	

11.5. a. Find the least square regression line for the following set of data:

{(−1, 0),(0, 2),(1, 4),(2, 5)}

b. Plot the given points and the regression line in the same rectangular system of axes.

11.6. The sales of a company (in million dollars) for each year are shown in Table 11.16.

TABLE 11.16   Company Sales for Exercise 11.6

X (year)	2005	2006	2007	2008	2009
Y (sales)	12	19	29	37	45

a. Find the least square regression line $y = ax + b$.

b. Use the least squares regression line as a model to estimate the sales of the company in 2012.

11.7. For the data on student and books:

Semester	Students	Books
1	36	31
2	28	29
3	35	34
4	39	35
5	30	29
6	30	30
7	31	30
8	38	38
9	36	34
10	38	33
11	29	29
12	26	26

a. Obtain a scatter plot of the number of books sold versus the number of registered students.

b. Find the correlation coefficient and interpret it in terms of this problem.

c. Give the regression equation and interpret the coefficients in terms of this problem.

d. If appropriate, predict the number of books that would be sold in a semester when 30 students have registered.

11.8. You must examine the relationship between the age and price for used cars sold in the last year by a car dealership company (Table 11.17).

TABLE 11.17   Age and Price Data for Exercise 11.8

Car Age (in years)	Price (in dollars)
4	6,300
4	5,800
5	5,700
5	4,500
7	4,500
7	4,200
8	4,100
9	3,100
10	2,100
11	2,500
12	2,200

a. Determine the correlation coefficient.

b. Build a simple linear model and interpret its diagnostics.

c. Build a quadratic model and interpret its diagnostics.

11.9. The time x in years that an employee spent at a company and the employee's hourly pay, y, for five employees are listed in Table 11.18.

a. Calculate and interpret the correlation coefficient. Include a plot of the data in your discussion.

b. Build a simple linear model.

c. Interpret the results.

TABLE 11.18 Employee Years and Pay for Exercise 11.9

x	y
5	25
3	20
4	21
10	35
15	38

11.10. This dataset of size $n = 15$ contains measurements of yield from an experiment done at five different temperature levels. The variables are y = yield and x = temperature in degrees Fahrenheit. Table 11.19 gives the data used for this analysis.

TABLE 11.19 Temperature and Yield for Exercise 11.10

i	Temperature	Yield
1	50	3.30
2	50	2.80
3	50	2.90
4	70	2.30
5	70	2.60
6	70	2.10
7	80	2.50
8	80	2.90
9	80	2.40
10	90	3.00
11	90	3.10
12	90	2.80
13	100	3.30
14	100	3.50
15	100	3.00

a. Determine the correlation coefficient.

b. Build a simple linear model and interpret its diagnostics.

c. Build a quadratic model and interpret its diagnostics.

REFERENCES AND SUGGESTED READINGS

Atmospheric carbon dioxide record from Mauna Loa: http://cdiac.esd.ornl.gov/trends/co$_2$/sio-mlo.htm

Atmospheric CO_2 concentrations (ppmv) derived from in situ air samples collected at Mauna Loa observatory, Hawaii: http://cdiac.esd.ornl.gov/ftp/maunaloa-co$_2$/maunaloa.co$_2$

Carbon dioxide information analysis center: http://cdiac.esd.ornl.gov

CHAPTER 12

Marketing Strategies and Competition Using Game Theory

I N STUDYING AND ANALYZING situations where your competitors' actions influence the outcomes to your company, you should consider employing game theory. We present examples of both total conflict games, also known as zero-sum or constant sum games, and partial conflict games.

12.1 TOTAL CONFLICT GAMES

Consider a situation where two companies, such as A and B, are competing in two regions for the market share. Each company may advertise or not advertise.

12.1.1 Market Shares

Suppose our marketing experts have analyzed a year's worth of data and have the following simplified payoff matrix for the percentage of the market share gained.

First, we note that the same of the pairs is always equal to zero. This makes this game a zero-sum game under the Total Conflict situation. As such we can simplify the payoff matrix by looking only at player A's results (Table 12.2).

 DOI: 10.1201/9781003464969-12

TABLE 12.1   Possible Payoff Matrix

		B	
		ADV	No ADV
A	ADV	50, −50	75, −75
	No ADV	−75, 75	60, −60

TABLE 12.2   Payoff matrix from A's perspective

		B	
		ADV	No ADV
A	ADV	50	75
	No ADV	25	60

We examine solutions by first looking for pure strategy solutions. If none exists, then we look for mixed strategy solutions. First, we define Nash Equilibrium that is the solution in Total Conflict Games.

Nash Equilibrium: A Nash Equilibrium is a set of outcomes where neither player can benefit by departing unilaterally from their strategy associated with that outcome.

How can we find this Nash Equilibrium? There are three methods that can be employed to find the Nash Equilibrium:

(1) Movement diagram

(2) Saddle points

(3) Linear Programming method

12.1.1.1 Movement Diagram

In a movement diagram with only one player's payoffs (zero-sum game) in each row, an arrow is drawn from the larger to the smaller value (with a tie the arrow goes both ways), and in each column, an arrow is drawn from the smaller to larger value.

In our example, the movement diagram looks like this (Figure 12.1):

We find all rows point to Strategies (ADV, ADV), which is our pure strategy solution.

		B		
		ADV	No ADV	
A	ADV	↑ 50 ⟵	75 ↑	
	No ADV	⎸ 25 ⟵	60 ⎸	

FIGURE 12.1   Movement Diagram.

12.1.1.2 Saddle Point

TABLE 12.3   Identifying Strategy Solution

		ADV	No ADV	Row Minimums	Max of Row Minimums
A	ADV	50	75	50	50
	No ADV	25	60	25	
	Col Max	50	75		
	Min of Col Max	50			

When the Max of Row Minimums equals the Min of Col Maximums, then we have a saddle point solution, which is a pure strategy solution. In this case, we have 50. Both companies should ADV and share the market here.

12.1.1.3 Linear Programming

Using the same generalizable conventions as in the Straffin text (2004), we call the row player, Rose, and the column player, Colin. Let's define the zero-sum game with the following payoff matrix that has components for both Rose and Colin where Rose has m strategies and Colin has n strategies:

$$(M,N) = \begin{bmatrix} (M_{1,1},N_{1,1}) & (M_{1,2},N_{1,2}) & \cdot & \cdot & \cdot & (M_{1,n},N_{1,n}) \\ (M_{2,1},N_{2,1}) & (M_{2,2},N_{2,2}) & \cdot & \cdot & \cdot & (M_{2,n},N_{2,n}) \\ \cdot & & \cdot & \cdot & \cdot & \cdot \\ \cdot & & \cdot & \cdot & \cdot & \cdot \\ \cdot & & \cdot & \cdot & \cdot & \cdot \\ (M_{m,1},N_{m,1}) & (M_{m,2},N_{m,2}) & \cdot & \cdot & \cdot & (M_{m,n},N_{m,n}) \end{bmatrix}$$

FIGURE 12.2   Payoff Matrix with n Strategies.

In the special case of zero-sum games, each pair sums to zero. For example, one such pair is $M_{11}+N_{11}= 0$. In the special case of the constant sum game, all pairs sum to the same constant, C. For example, the sum of all $M_{ij} + N_{ij} = C$.

Our knowledge of the zero-sum game and the primal–dual relations suggests formulating Rose's game and finding Colin's solution through the dual solution. This works well if you have a single problem to solve. But what if you have many games to consider and you only have Excel. How best to construct a technology assistant?

Treating the zero-sum game as above translates into two linear programming formulations, one for each maximizing player. We combine the formulation into a single formulation shown in Equation (12.3) where the solution provides the values of the game and the probabilities that the players should play their strategies. Further, if any pairs $(M_{mn}, N_{m,n})$ are negative, then there is a chance that the game solution can be negative. Since the game solution will be a decision variable in our formulation, we must account for that possibility. Our best recommendation is to use the method suggested by Winston [1995, pp. 172–178] to replace any variable that could take on negative values with the difference in two positive variables, $x_j - x'_j$. We assume that the value of the game could be positive or negative. The other values we are looking for are probabilities that are always between 0 and 1. Since this occurs only in the value of the game, we use as a substitute variable, $V = v_1 - v_2$.

Objective function:

$$\text{Maximize } v_1 - v2 \qquad \text{(Eq. 12.1)}$$

Subject to:

$$M_{1,1}x_1 + M_{2,1}x_2 + \ldots + M_{m,1}x_n - v_1 + v_2 \geq 0$$
$$M_{1,2}x_1 + M_{2,2}x_2 + \ldots + M_{m,2,}x_n - v_1 + v_2 \geq 0$$
$$\ldots$$
$$M_{1,m}x_1 + M_{2,m}x_2 + \ldots + M_{m,n}x_n - v_1 + v_2 \geq 0$$
$$x_1 + x_2 + \ldots + x_n = 1$$
$$Nonnegativity$$

where the weights x_i yield Rose strategy, and the value of V is the value of the game to Rose.

Objective function:

$$\text{Maximize } v_3 {}_- v_4 \tag{Eq. 12.2}$$

Subject to:

$$N_{1,1}y_1 + N_{1,2}y_2 + \ldots + N_{1,m}y_n - v_3 + v_4 \geq 0$$
$$N_{2,1}y_1 + N_{2,2}y_2 + \ldots + N_{2,m}y_n - v_3 + v_4 \geq 0$$

...

$$N_{m,1}y_1 + N_{m,2}y_2 + \ldots + N_{m,n}y_n - v_3 + v_4 \geq 0$$
$$y_1 + y_2 + \ldots + y_n = 1$$

Nonnegativity

where the weights y_i yield Colin's strategy, and the value of $v_3 - v_4$ is the value of the game to Colin.

To accomplish this as one formulation, we combine as

Objective function:

$$\text{Maximize } v_1 - v_2 + v_3 - v_4 \tag{Eq. 12.3}$$

Subject to:

$$M_{1,1}x_1 + M_{2,1}x_2 + \ldots + M_{m,1}x_n - v_1 + v_2 \geq 0$$
$$M_{1,2}x_1 + M_{2,2}x_2 + \ldots + M_{m,2}x_n - v_1 + v_2 \geq 0$$

...

$$M_{1,m}x_1 + M_{2,m}x_2 + \ldots + M_{m,n}x_n - v_1 + v_2 \geq 0$$
$$x_1 + x_2 + \ldots + x_n = 1$$

Nonnegativity

$$N_{1,1}y_1 + N_{1,2}y_2 + \ldots + N_{1,m}y_n - v_3 + v_4 \geq 0$$
$$N_{2,1}y_1 + N_{2,2}y_2 + \ldots + N_{2,m}y_n - v_3 + v_4 \geq 0$$

...

$$N_{m,1}y_1 + N_{m,2}y_2 + \ldots + N_{m,n}y_n - v_3 + v_4 \geq 0$$
$$y_1 + y_2 + \ldots + y_n = 1$$

Nonnegativity

In our example, since we have no negative values as payoff entries, we can use just the value of the game in lieu of $v_1 - v_2$.

Let P = Payoff to the row player as the value of the game

Let

$$x_1 = ADV$$

$$x_2 = No\ ADV$$

Objective function: Max P
   Subject to:

$$50\,x_1 + 25\,x_2 - P \geq 0$$

$$75\,x_1 + 60\,x_2 - P \geq 0$$

$$x_1 + x_2 = 1$$

All variables are nonnegative.
   LINDO solution:

LP OPTIMUM FOUND AT STEP 2

   OBJECTIVE FUNCTION VALUE

   1)   50.00000

VARIABLE	VALUE	REDUCED COST
P	50.000000	0.000000
X_1	1.000000	0.000000
X_2	0.000000	25.000000

ROW	SLACK OR SURPLUS	DUAL PRICES
2)	0.000000	-1.000000
3)	25.000000	0.000000
4)	0.000000	50.000000

NO. OF ITERATIONS= 2

RANGES IN WHICH THE BASIS IS UNCHANGED:

   OBJ COEFFICIENT RANGES

VARIABLE	CURRENT	ALLOWABLE	ALLOWABLE
	COEF	INCREASE	DECREASE
P	1.000000	INFINITY	1.000000

```
      X₁      0.000000                    INFINITY  25.000000
      X₂      0.000000                    25.000000  INFINITY


    RIGHT-HAND SIDE RANGES

  ROW          CURRENT         ALLOWABLE        ALLOWABLE
  RHS          INCREASE        DECREASE
   2           0.000000        50.000000        25.000000
   3           0.000000        25.000000        INFINITY
   4           1.000000        INFINITY         1.000000
```

We change scenarios to illustrate a solution by Nash Equilibrium that is a mixed strategy.

12.1.2 Hitter–Pitcher Dual – A Conflict Game Example

Movement diagram is shown in Figure 12.3.

We find the arrows do not point to a strategy; they move around.

Saddle point solution is given in Table 12.5.

Here, $0.295 \neq 0.250$; so we do not have a saddle point solution. Either method that produces no pure strategy solution is sufficient so we can move on to a mixed strategy solution.

Mixed strategy: If each player only has two strategies, as is the case here, we can use the Method of Oddment, otherwise we recommend Linear Programming.

Oddments:

The mixed strategy (Nash Equilibrium – Figure 12.4) is for the hitter to guess fastball 48.72% of the time and guess slider 51.82% of the time and for the pitcher to through a fastball 23.07% of the time and the slider 76.93% of the time. The payoff is:

$$0.4872(350) + 0.5182 (0.2) = 0.27416$$

Linear Programming:

Objective function: MAX BA

Subject to:

$$- BA + 0.35 \, X_1 + 0.2 \, X_2 \geq 0$$

$$- BA + 0.25 \, X_1 + 0.295 \, X_2 \geq 0$$

TABLE 12.4   Payoff Matrix from a Hitter–Pitcher Dual

		Pitcher	
	Guess/Pitch	Fastball	Slider
Hitter	Fastball	0.35	0.25
	Slider	0.20	0.295

		Pitcher	
	Guess/Pitch	Fastball	Slider
Hitter	Fastball	↑0.35➡	0.25 ↓
	Slider	↑0.20⬅	0.295↓

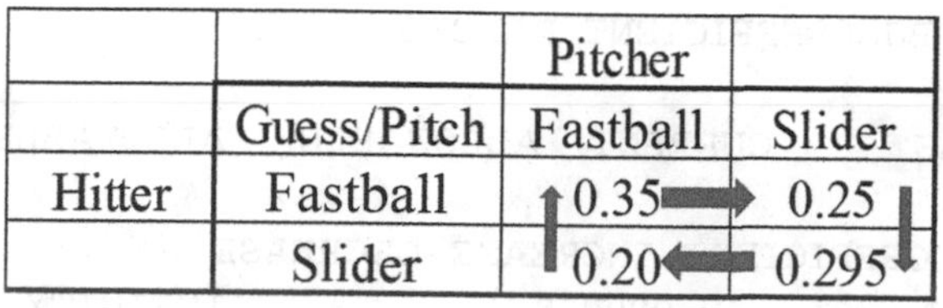

FIGURE 12.3   Movement Diagram.

TABLE 12.5   Strategy for Hitter–Pitcher Dual.

		Pitcher			
	Guess/Pitch	Fastball	Slider	Row mins	Max of Row Min
Hitter	Fastball	0.35	0.250	0.25	0.25
	Slider	0.20	0.295	0.20	
	Col Max	0.35	0.295		
	Min of Col Max		0.295		

$$X_1 + X_2 = 1$$

END

LP OPTIMUM FOUND AT STEP 2

OBJECTIVE FUNCTION VALUE

1)   0.2730769

VARIABLE	VALUE	REDUCED COST
BA	0.273077	0.000000
X_1	0.487179	0.000000
X_2	0.512821	0.000000

ROW   SLACK OR SURPLUS   DUAL PRICES

2)	0.000000	−0.230769
3)	0.000000	−0.769231
4)	0.000000	0.273077

NO. OF ITERATIONS = 2

RANGES IN WHICH THE BASIS IS UNCHANGED:

OBJ COEFFICIENT RANGES

VARIABLE	CURRENT COEFFICIENT	ALLOWABLE INCREASE	ALLOWABLE DECREASE
BA	1.000000	INFINITY	1.000000
X_1	0.000000	0.045000	0.150000
X_2	0.000000	0.150000	0.045000

RIGHT-HAND SIDE RANGES

ROW RHS	CURRENT INCREASE	ALLOWABLE	ALLOWABLE DECREASE
2	0.000000	0.100000	0.095000
3	0.000000	0.095000	0.100000
4	1.000000	INFINITY	1.000000

We note that the minor difference in the objective value is due to the rounding of the probabilities used in the mixed strategies. If more than two strategies are available to each player, then we recommend the LP approach.

12.1.3 The Expanded Hitter–Pitcher Dual

In view of the use of technology in sports today, we present an example of the hitter–pitcher duel. First in this example we extend the strategies for each player in our model. We consider a batter-pitcher duel between a hitter of the Philadelphia Phillies and various pitchers in the national league where the pitcher throws a fastball, a split-finger fastball, a curve ball, and a changeup. The batter, aware of these pitches, must prepare appropriately for the pitch. Data is available from many websites that we might use. In this example, we obtained the data from the internet, www.STATS.com. We consider both a right-handed pitcher and a left-handed pitcher separately in this analysis.

	Guess/Pitch	Pitcher			
		Fastball	Slider	Fastball-Slider or Slider-Fastball, whichever is > 0	Mixed Strategies
Hitter	Fastball	0.35	0.250	0.100	.095/.195 = 0.4872
	Slider	0.20	0.295	0.095	.1/.195 = 0.5182
		0.15	0.045	Sum = 0.1095	
		.045/.195 = 0.2307	.150/.195 = 0.7692		

FIGURE 12.4   Strategy for the Hitter–Pitcher Problem.

TABLE 12.6   Baseball Data for Expanded Hitter–Pitcher Dual

Hitter/RHP	FB	CB	CH	SF
FB	0.337	0.246	0.220	0.200
CB	0.283	0.571	0.339	0.303
CH	0.188	0.347	0.714	0.227
SF	0.200	0.227	0.154	0.500

For a National League right-handed pitcher (RHP) versus Ryan Howard, we have compiled the following data (Table 12.6). Let FB = fastball, CB = curveball, CH = changeup, SF = split-fingered fastball.

Both the batter and pitcher want the best possible result. We set this up as a linear programming problem. Our decision variables are x_1, x_2, x_3, and x_4 as the percentages to guess FB, CB, CH, and SF, respectively, and $\underline{V}$ represents the hitter's batting average.

Objective function:

Max V

Subject to:

$$0.337x_1 + 0.283x_2 + 0.188x_3 + 0.200x_4 - V \geq 0$$

$$0.246x_1 + 0.571x_2 + 0.347x_3 + 0.227x_4 - V \geq 0$$

$$0.220x_1 + 0.339x_2 + 0.714x_3 + 0.154x_4 - V \geq 0$$

$$0.200x_1 + 0.303x_2 + 0.227x_3 + 0.500x_4 - V \geq 0$$

$$x_1 + x_2 + x_3 + x_4 = 1$$

$$x_1, x_2, x_3, x_4, V \geq 0$$

We solve this linear programming problem and find the optimal solution (strategy) is to guess the fastball (FB) 27.49%, guess the curveball (CB)

64.23%, never guess changeup (CH), and guess split-finger fastball (SF) 8.27% of the time to obtain a 0.291 batting average.

The pitcher then wants to also keep the batting average as low as possible. We can set up the linear program for the pitcher as follows.

Our decision variables are y_1, y_2, y_3, and y_4 as the percentages to throw the FB, CB, CH, and SF respectively, and V represents the hitter's batting average.

Objective function: Min V

Subject to:

$$0.337y_1 + 0.246y_2 + 0.220y_3 + 0.200y_4 - V \leq 0$$

$$0.283y_1 + 0.571y_2 + 0.339y_3 + 0.303y_4 - V \leq 0$$

$$0.188y_1 + 0.347y_2 + 0.714y_3 + 0.227y_4 - V \leq 0$$

$$0.200y_1 + 0.227y_2 + 0.154y_3 + 0.500y_4 - V \leq 0$$

$$y_1 + y_2 + y_3 + y_4 = 1$$

$$x_1, y_2, y_3, y_4, V \geq 0$$

We find the RHP should randomly throw 65.93% fastballs, no curveballs, 3.25% changeups, and 30.82% split-finger fastballs for Howard to keep only the 0.291 batting average.

12.2 THE PARTIAL CONFLICT GAME ANALYSIS WITHOUT COMMUNICATION

Let's first define a partial conflict game. As opposed to a total conflict game where if a player wins x his opponent loses x, in a partial conflict game the players are not strictly opposed, so it is possible for both players to win or lose some value. In a partial sum game, the sum of the values for the two players do not sum to zero.

In partial conflict games, the Nash Equilibrium may not be the solution to the game. There are many strategies to try and analysis to do to find the solution.

12.2.1 Example 12.1

Consider the following game partial conflict game (Table 12.7) where the sums of the outcomes do not all sum to zero or the same constant.

TABLE 12.7   2 × 2 Partial Conflict Game

		Player II	
		C_1	C_2
Player I	R_1	(2, 4)	(1, 0)
	R_2	(3, 1)	(0, 4)

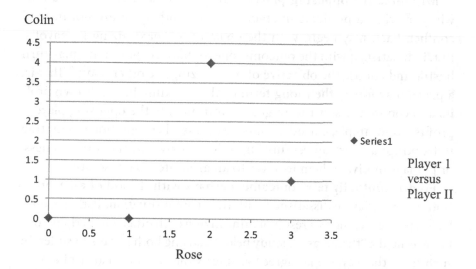

FIGURE 12.5   Payoffs in a Partial Conflict Game Do Not Lie on a Line.

In Figure 12.5, we note that a plot of the payoffs to each player does not lie on a line, indicating that the game is a partial conflict game because total conflict game values lie in a straight line.

What are the objectives of the players in a partial conflict game? In total conflict, each player attempts to maximize his payoffs and necessarily minimizes the other player in the process. But in a partial conflict game, a player may have any of the following objectives from Giordano et al. (2014).

Maximize his payoffs: Each player chooses a strategy in an attempt to maximize his payoff. While he reasons what the other player's response will be, he does not have the objective of ensuring the other player gets a "fair" outcome. Instead, he "selfishly" maximizes his payoff.

Find a stable outcome: Quite often, players have an interest in finding a stable outcome. *A Nash Equilibrium outcome is an outcome from which* *neither player can unilaterally improve* and therefore represents a stable situation. For example, we may be interested in determining whether two species in a habitat will find equilibrium and coexist, or will one species dominate and drive the other to extinction? The Nash Equilibrium is named in honor of John Nash (1950) who proved that every two-person game has at least one equilibrium in either pure strategies or mixed strategies.

Minimize the opposing player: Suppose we have two corporations whose marketed products interact with each other but are not in total conflict. Each may begin with the objective of maximizing its payoffs. But, if dissatisfied with the outcome, one, or both corporations, may turn hostile and choose the objective of minimizing the other player. That is, a player may forego their long-term goal of maximizing their own profits and choose the short-term goal of minimizing the opposing player's profits. For example, consider a large, successful corporation attempting to bankrupt a "start-up venture" in order to drive them out of business, or perhaps motivate them to agree to an arbitrated "fair" solution.

Find a "mutually fair" outcome, perhaps with the aid of an arbiter: Both players may be dissatisfied with the current situation. Perhaps, both have a poor outcome as a result of minimizing each other. Or, perhaps, one has executed a "threat' as we study below, causing both players to suffer. In such cases, the players may agree to abide by the decision of an arbiter who must then determine a "fair" solution.

In this introduction to partial conflict games, we will assume that both players have the objective of maximizing their payoffs. Next, we must determine if the game is played without communication or with communication. "Without communication" indicates that the players must choose their strategies without knowing the choice of the opposing player. For example, perhaps they choose their strategies simultaneously. The term "with communication" indicates that perhaps one player can move first and make his move known to the other player, or that the players can talk to one another before they move. We assume that our games do not allow communication and are played simultaneously.

Further, we assume our players are rational, attempting to obtain their best outcomes and that games are repetitive.

One method to find a pure strategy solution is the movement diagram.

Movement diagram: For Player I, examine the first value in the coordinate and compare R_1 to $\underline{R}_2$. For each C_1 and C_2, draw an arrow from the smaller to larger values between R_1 and R_2. For Player II, examine the second value in the coordinate and compare C_1 to C_2. For each R_1 and R_2, draw an arrow from the smaller to larger values between C_1 and C_2.

For example, under C_1, we draw an arrow from 2 to 3 and under C_2 from 0 to 1. Under R_1, we draw the arrow from 0 to 4 and under R_2 from 1 to 4. We show this in Figure 12.6.

Using the Excel template, Figure 12.7 the arrows indicate "false" in all directions so there is no pure strategy.

We follow the arrows. If the arrows lead us to a value or values where no arrows point out, then we have a pure strategy solution. If the arrows move in a clockwise or counterclockwise direction, then we have no pure strategy solution. Here, we move counterclockwise and have no pure strategy solution. Nash proved that all games have a solution either by pure or mixed strategies. As a matter of fact, Barron (2013) showed that some partial conflict games have both a pure and mixed (equalizing) strategy.

We start here by defining the mixed (equalizing) strategy for a partial conflict game.

Rose's game: Rose maximizing, Colin "equalizing" is a total conflict game that yields Colin's equalizing strategy.

Colin's game: Colin maximizing, Rose "equalizing" is a total conflict game that yields Rose's equalizing strategy.

		Player II	
	C_1		C_2
R_1	(2, 4)	⟵	(1, 0)
Player I	↓		↑
R_2	(3, 1)	⟶	(0, 4)

FIGURE 12.6 Movement Diagram.

For Solving a 2 X 2 game for Equalizing Strategies
Step 1. Enter Rose's and Colin's Values into the appropriate cells

			Colin			
		C_1			C_2	
	R_1	2	4 <--	1	0	
Rose		down_|/		UP_/|		
	R_2	3	1 -->	0	4	
Follow the arrows:		FALSE			0	
		FALSE			0	

FIGURE 12.7 Screenshot of Excel Template for Movement Diagram.

Note: If either side plays its equalizing strategy, then the other side "unilaterally" cannot improve its own situation (it stymies the other player).

We will call this strategy an equalizing strategy. Each player is restricting what his opponent can obtain by ensuring no matter what they do that his opponent always gets the identical solution (Straffin, 2004).

12.3 METHODS TO OBTAIN THE EQUALIZING STRATEGIES

We present two methods to obtain equalizing strategies, and we will apply these methods to our previous example. The two methods are: Linear programming and nonlinear programming. We state here that linear programming works only because each player has only two strategies.

12.3.1 Linear Programming with Two Players and Two Strategies Each

This translates into two maximizing linear programming formulations as shown in Equations 12.1 and 12.2. Formulation (12.4) provides the Nash Equalizing solution for Colin with strategies played by Rose, while formulation (12.3) provides the Nash Equalizing solution for Rose and strategies played by Colin. The two constraints representing strategies are implicitly equal to each other as per this formulation (see Fox, 2010).

Objective function: Maximize V
Subject to:

$$N_{1,1}x_1 + N_{2,1}x_2 - V \geq 0$$
$$N_{1,2}x_1 + N_{2,2}x_2 - V \geq 0$$
$$(N_{1,1} - N_{1,2})x_1 + (N_{2,1} - N_{2,2})x_2 = 0 \qquad \text{(Eq. 12.4)}$$
$$x_1 + x_2 = 1$$
$$Nonnegativity$$

Objective function: Maximize v

Subject to:

$$M_{1,1}y_1 + M_{1,2}y_2 - v \geq 0$$
$$M_{2,1}y_1 + M_{2,2}y_2 - v \geq 0$$
$$(M_{1,1} - M_{2,1})y_1 + (M_{1,2} - M_{2,2})y_2 = 0 \qquad \text{(Eq. 12.5)}$$
$$y_1 + y_2 = 1$$
$$Nonnegativity$$

With our example, we obtain the following formulation:

Objective function: Maximize V

Subject to:

$$4x_1 + x_2 - V \geq 0$$
$$0x_1 + 4x_2 - V \geq 0$$
$$4x_1 - 3x_2 = 0$$
$$x_1 + x_2 = 1$$
$$Nonnegativity$$

and

Objective function: Maximize v

Subject to:

$$2y_1 + y_2 - v \geq 0$$
$$3y_1 + 0y_2 - v \geq 0$$
$$-y_1 + y_2 = 0$$
$$y_1 + y_2 = 1$$
$$Nonnegativity$$

12.3.1.1 Excel to Obtain Equalizing Strategies

We will now use Excel's Solver to address the two players and two-strategy problem (Figure 12.8).

$3/7 \, x_1$, $4/7 \, x_2$ corresponding to $3/7 \, R_1$, $4/7 \, R_2$ and $\frac{1}{2} \, y_1$, $\frac{1}{2} \, y_2$ corresponding to $\frac{1}{2} \, C_1$, $\frac{1}{2} \, C_2$. The Nash Equilibrium is (3/2, 16/7).

12.3.1.2 Nonlinear Programming Approach for Two or More Strategies for Each Player

For games with two players and more than two strategies each, we present the nonlinear optimization approach by Barron (2013). Consider a two-person game with a payoff matrix as before. Let's separate the payoff matrix into two matrices **M** and **N** for Players I and II. We solve the following nonlinear optimization formulation in expanded form.

Objective function:

$$Maximize \sum_{i=1}^{n} \sum_{j=1}^{m} x_i a_{ij} y_j + \sum_{i=1}^{n} \sum_{j=1}^{m} x_i b_{ij} y_j + -p - q$$

	Linear Programming					
Decision Variables						
x_1	0.571429					
x_2	0.428571					
vc	1.714286					
y_1	0.5					
y_2	0.5					
vr	1.5					
OBJ	3.214286					
Constraints			0	0		$2y_1 +y_2 -vr>=0$
			0	0		$3*y_1 -vc>=0$
			1	1		$y_1 +y_2 =1$
			1	0		$4x_1 +x_2 -vc>=0$
			0	0		$4x_2 -vc>=0$
			1	1		$x_1 +x_2 =1$
			0	0		$-y_1 +y_2 =0'$
			0	0		$3x_1 -4x_2 =0$

FIGURE 12.8 Screenshot of Excel for Two-Player Strategy.

Subject to:

$$\sum_{j=1}^{m} a_{ij}y_j \le p, \quad i=1,2,\ldots,n,$$

$$\sum_{i=1}^{n} x_i b_{ij} \le q, \quad j=1,2,\ldots,m,$$

$$\sum_{i=1}^{n} x_i = \sum_{j=1}^{m} y_j = 1$$

$$x_i \ge 0, y_j \ge 0$$

We return to our previous example. We define M and N as:

$$M = \begin{bmatrix} 2 & 1 \\ 3 & 0 \end{bmatrix} and \ N = \begin{bmatrix} 4 & 0 \\ 1 & 4 \end{bmatrix}$$

We define x_1, x_2, y_1, y_2 as the probabilities for players playing their respective strategies.

By substitution and simplification, we obtain the following:
Objective function:

Maximize

$$6y_1x_1 + 4y_1x_2 + x_1y_2 + 4x_2y_2 - p - q$$

Subject to:

$$x_1 + x_2 = 1$$

$$y_1 + y_2 = 1$$

$$4x_2 - q \leq 0$$

$$4x_1 + x_2 - q \leq 0$$

$$2y_1 + y_2 - p \leq 0$$

$$3y_1 - p \leq 0$$

Non-negativity

We find the exact same solution as before with the larger screenshot (Figure 12.9).

12.3.1.3 Finding a Solution

According to Straffin (2004), a Nash equilibrium is a solution if and only if it is unique and Pareto optimal. Pareto optimality refers to the northeast region of a payoff polygon where the payoff polygon is found as the convex set formed by the outcome coordinates (Figure 12.10).

We see in the figure that the Nash equilibrium (1.5, 2.28) is not Pareto optimal (see Figure 12.11) and not the solution that we should seek.

At this point, we might try to allow communication and try strategic moves which we do not describe here but can be reviewed in Giordano et al. (2014). Further, we might want to show the method of Nash arbitration although we do not illustrate that here.

12.4 NASH ARBITRATION METHOD

When we have not achieved an acceptable Pareto-optimal solution by other methods that are acceptable by the players, then a game might move to arbitration. The Nash arbitration theorem (1950) states that,

A				
	2	1		
	3	0		
B	4	0		
	1	4		
dv				
x_1	0.428572			
x_2	0.571429			
p	1.500001			
y_1	0.5			
y_2	0.5			
q	2.285715			

FIGURE 12.9   Excel Screenshot of Nonlinear Approach for Two-Player Strategy.

there is one and only arbitration scheme which satisfies rational-ity, linear invariance, symmetry, and independence of irrelevant alternatives. It is this: if the status quo (SQ) point is (x_0, y_0), then the arbitrated solution point N is the point (x, y) in the polygon with $x \geq x_0$, $y \geq y_0$ which maximizes the product of $(x-x_0)(y-y_0)$.

There are a few terms, strategies, and methods we must discuss prior to illustrating this entire process. Using this security level as our status quo point, we now formulate the Nash arbitration scheme. There are four axioms that are required to be met using the arbitration scheme.

- Axiom 1: Rationality. The solution should be in the negotiation set.

- Axiom 2: Linear Invariance. If either Rose's or Colin's utilities are transformed by a positive linear function, the solution point should be transformed by the same function.

	A	B	C	D	E	F	G
	Decision Variables						
	x_1	0.428571				Objective Function	
	x_2	0.571429				z	0
	x_3	0					
	x_4	0					
	x_5	0			X^TAY^T	1.5	
	x_6	0			X^TBY^T	2.285714	
	x_7	0					
	x_8	0					
	x_9	0					
	x_{10}	0					
	p	1.5					
	v_2	0					
	y_1	0.5					
	y_2	0.5					
	y_3	0					
	y_4	0					
	y_5	0					
	y_6	0					
	y_7	0					
	y_8	0					
	y_9	0					
	y_{10}	0					
	q	2.285714					
	v_4	0					

FIGURE 12.10 Excel Screenshot of Nonlinear Approach for Two-Player Strategy.

- Axiom 3. Symmetry. If the polygon happens to be symmetric about the line of slope +1 through the status quo point, then the solution should be on this line.

- Axiom 4: Independence of Irrelevant Alternatives. Suppose N is the solution point for a polygon, P with status quo point SQ. Suppose Q is another polygon which contains both SQ and N and is totally contained in P. Then N should also be the solution point to Q with status quo point SQ.

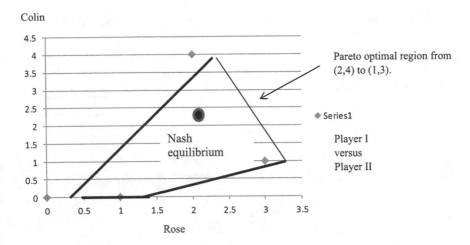

FIGURE 12.11   Payoff Polygon and Pareto-Optimal Region.

- First, we will define the SQ point as either the security levels found through prudential strategies or the threat level found by communications methods. We will use only the security levels here.

12.4.1 Finding the Prudential Strategy (Security Levels)

The security levels are the payoffs to the players in a partial conflict game where each player attempts to maximize their own payoff. We can solve for these payoffs using a separate linear program for each security level.

We have Rose in Rose's game and Colin in Colin's game. So, the LP formulation is:

Rose objective function:

Maximize V

Subject to:

$$N_{1,1}y_1 + N_{1,2}y_2 + \ldots + N_{1,m}y_n - V \geq 0$$
$$N_{2,1}y_1 + N_{2,2}y_2 + \ldots + N_{2,m}y_n - V \geq 0$$
$$\ldots$$
$$N_{m,1}x_1 + N_{m,2}x_2 + \ldots + N_{m,n}x_n - V \geq 0$$
$$y_1 + y_2 + \ldots + y_n = 1$$
$$y_j \leq 1 \quad for \quad j = 1,\ldots,n$$
$$Nonnegativity$$

where the weights y_i yield Colin's prudential strategy and the value of V is the security level for Colin objective function:

Maximize v

Subject to:

$$M_{1,1}x_1 + M_{2,1}x_2 + \ldots + M_{n,1}x_n - v \geq 0$$
$$M_{1,2}x_1 + M_{2,2}x_2 + \ldots + M_{n,2}x_n - v \geq 0$$
$$\ldots$$
$$M_{1,m}y_1 + M_{2,m}y_2 + \ldots + M_{m,n}y_n - v \geq 0$$
$$x_1 + x_2 + \ldots + x_m = 1$$
$$x_i \leq 1 \quad for \quad i = 1,\ldots,m$$
$$Nonnegativity$$

12.4.2 Example 12.2

Given the following payoff matrix (Table 12.8) from Example 12.1, Section 12.3.

We plot and see the payoff polygon in Figure 12.12.

Let's return to Example 12.1 to illustrate finding the security levels. Let SLR and SLC represent the security levels for Rose and Colin, respectively. We use linear programming to find these values using the following formulations:
Objective function:

Max SLR

Subject to:

$$2x_1 + 3x_2 - SLR \geq 0$$

$$1x_1 + 0x_2 - SLR \geq 0$$

$$x_1 + x_2 = 1$$

TABLE 12.8   Payoff Matrix

		Player II	
		C_1	C_2
Player I	R_1	(2, 4)	(1, 0)
	R_2	(3, 1)	(0, 4)

Colin

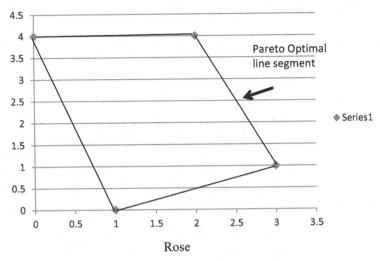

FIGURE 12.12   Payoff Polygon and Pareto-Optimal Region.

Non-negativity

Objective function:

$$Max\ SLC$$

Subject to:

$$4y_1 - SLC \geq 0$$

$$1y_1 + 4y_2 - SLC \geq 0$$

$$y_1 + y_2 = 1$$

Non-negativity

The solution yields both how the game with prudential strategies is played and the security levels. Player I always plays R_1 and Player II plays 4/7 C_1 and 3/7 C_2. The security level, the value of each player in prudential strategy, is *(1, 16/7) = (1, 2.286)*.

Using this security level as our status quo point, we can now formulate the Nash arbitration scheme, see Figure 12.13.

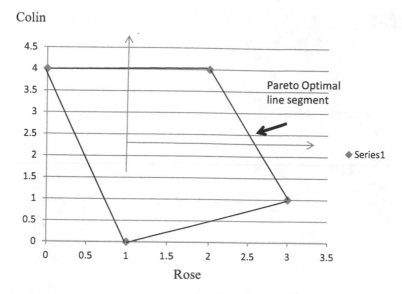

Colin

FIGURE 12.13   Payoff Polygon and Pareto-Optimal Region.

We apply this theorem in a nonlinear optimization model framework, Objective function:

$$Maximize\ (x - 1)(y - 2.286)$$

Subject to:

$$3x + y = 10$$
$$x \geq 1$$

$$y \geq 2.286$$

We find, in this example, that our Nash arbitration point for status quo (1, 2.286) is the point (2,4).

Here, we recommend the Solver with GRG nonlinear routine to find the solution. We provide screenshots of the input to the Solver and the resulting output solution (Figures 12.14 and 12.15).

After finding the Nash arbitration point, we should also determine how this point is obtained from the end points of the Pareto-optimal line segment. We will solve for the probabilities to play (2,4) and to play (3,1). Here is the formulation we will use:

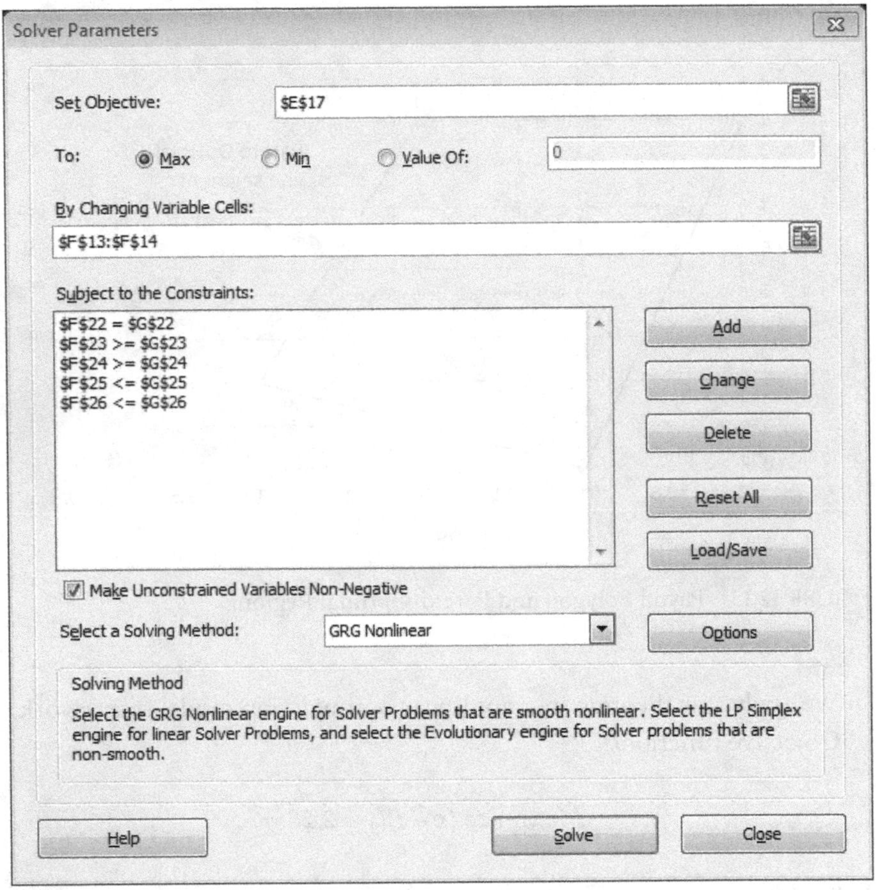

FIGURE 12.14   Excel Solver Screen for Nash Arbitration.

$$2p_1 + 3p_2 = 2 \; (x \text{ coordinate of Nash arbitration point})$$

$$4p_1 + 1p_2 = 4 \; (y \text{ coordinate of Nash arbitration point})$$

This solves with $p_1 = 1$, $p_2 = 0$. The arbitrators hold on to the strategies that yield (2,4) always.

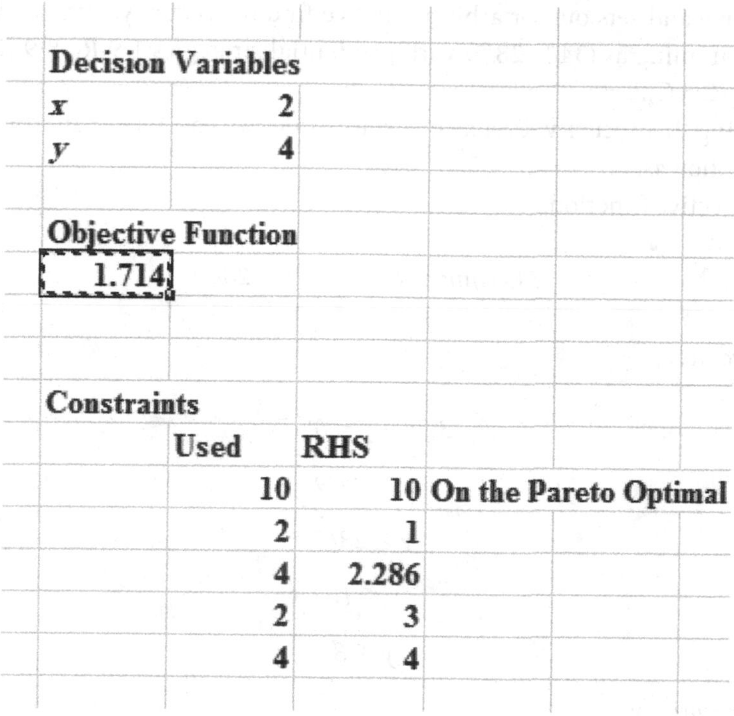

Decision Variables			
x	2		
y	4		
Objective Function			
1.714			
Constraints			
	Used	RHS	
	10	10	On the Pareto Optimal
	2	1	
	4	2.286	
	2	3	
	4	4	

FIGURE 12.15 Excel Screenshot of Solution for Nash Arbitration.

12.4.3 Example 12.3

Consider the payoff matrix (Table 12.9).

TABLE 12.9 Payoff Matrix.

		Colin	
		C_1	C_2
Rose	R_1	(2, 6)	(10, 5)
	R_2	(4, 4)	(3, 8)

The Nash equilibrium is found through equalizing strategies as (34/9, 28/5). It is not Pareto optimal, so Rose is assumed to be unhappy with her outcome and sets out for arbitration. We find the security levels by linear programming as (34/9, 28/5) with prudential strategies 1/9 R_1, 8/9 R_2 and 3/5 C_1, 2/5 C_2.

Using this security level as our status quo, we can formulate the Nash arbitration as:

Objective function:

$$Maximize\ (x - 34/9)(y - 28/5)$$

Subject to:

$$3/7x + y = 65/7$$

$$x \geq 34/9$$

$$y \geq 28/5$$

$$x \leq 10$$

$$y \leq 8$$

Nonnegativity

Our Nash-arbitrated solution (Figure 12.16) is (6.1889, 6.6333) which is better for Rose than the equilibrium value of 3.78 and better for Colin than his equilibrium value of 5.6. Both players are better off with arbitration. The scheme to achieve this output is found from the solution to the following systems of equations:

$$10p_1 + 3p_2 = 6.18889$$

$$5p_1 + 8p_2 = 6.63333$$

We find $p_1 = 45.5\%$ and $p_2 = 54.5\%$. Further, since both (10,5) and (3,8) are in C_2, Colin always plays C_2 while Rose plays 45.5 R_1 and 54.5% R_2.

12.4.4 Example 12.4: Writer's Guild Strike and Nash Arbitration

We examine strategic moves. The writers move first, and their best results is again (2,4). If management moves first, the best result is (2,4). First moves keep us at the Nash equilibrium. The writers consider a threat and tell management that if they choose SQ, they will strike putting us at (1,3).

Decision Variables						
x	6.18889					
y	6.63333					
Objective Function						
2.49148						
Constraints						
	Used	RHS				
	9.28572	9.28571	On the Pareto Optimal Line			
	6.18889	3.78				
	6.63333	5.6				
	6.18889	10				
	6.63333	8				

FIGURE 12.16   Excel Screenshot of Solution for Nash Arbitration.

This result is indeed a threat as it is worse for both the writers and management.

However, the options for management under IN (Increase to writers plus revenue sharing) are both worse than (1,3) so they do not accept the threat. The writers do not have a promise. At this point, we might involve an arbiter using the method as suggested earlier.

Writers and management security levels are found from prudential strategies and the following LP models. The security levels are (2, 3). We show this in Figure 12.16. The Nash arbitration formulation is:

Objective function:

$$\text{Maximize } (x - 2)(y - 3)$$

Subject to:

$$3/2\ x + y = 7$$

$$x \geq 2$$

$$y \geq 3$$

Colin

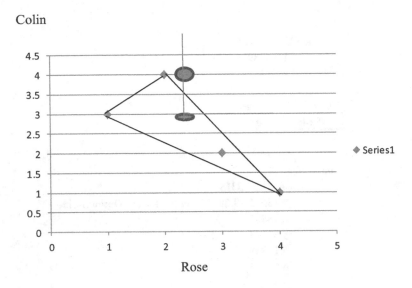

Rose

FIGURE 12.17   Payoff Polygon for Writer's Guild Strike.

The Nash equilibrium value, *(2,4)*, lies along the Pareto-optimal line segment (Figure 12.17). But the writers want to do better by going on strike and forcing arbitration, which is what they did. In this example, we consider "binding arbitration" where the players have a third party work out the outcomes that best meet their desires and are acceptable to all players. Nash found that this outcome can be obtained by the following steps:

The status quo point is the security levels of each side. We find these values using prudential strategies as (2,3). The function for the Nash Arbitration scheme is *Maximize (x – 2)(y – 3)*.

Using technology, we find the desired solution to our NLP as:

$$x = 2.3333$$

$$y = 3.5$$

We have the x and y coordinates *(2.3333, 3.5)* as our arbitrated solution. We can also determine how the arbiters should proceed. We solve the following simultaneous equations:

$$2p_1 + 4p_2 = 2.3333$$

$$4p_1 + p_2 = 3.5$$

We find that the probabilities to be played are 5/6 and 1/6. Further, we see that Player I, the writers, always play R_2 so the management arbiter plays 5/6 C_1 and 1/6 C_2 during the arbitration.

12.4.5 R and the Nash Arbitration Method

The CRAN library routines for game theory are beyond our coverage. However, we provide some R routines to find the Nash Arbitration point.

First, we need to have solved for the security values and have the function to maximize. We return to our writer's guild example.

Objective function:

$$\text{Maximize } (x - 2)\,(y - 3)$$

Subject to:

$$3/2\, x + y = 7$$

$$x \geq 2$$

$$y \geq 3$$

Recall, by default, R solves the minimum so we will use $-1 \times (x - 2) \times (y - 3)$ and the optimization methods described in Chapter 7.

First, we convert to a single variable problem and use Golden Section.

R Code

```
f = function(x)
+ {
+ -1* (x – 2)*(4–1.5*x)
+}
>
> golden.section.search(f,2,4,0.05)
```

R Output

Iteration # 1

$f_1 = 0.1114562$

$f_2 = 1.055728$

$f_2 > f_1$

New Upper Bound = 3.236068

New Lower Bound = 2

New Upper Test Point = 2.763932

New Lower Test Point = 2.472136

Iteration # 2

$f_1 = -0.1377674$

$f_2 = 0.1114562$

$f_2 > f_1$

New Upper Bound = 2.763932

New Lower Bound = 2

New Upper Test Point = 2.472136

New Lower Test Point = 2.291796

Iteration # 3

$f_1 = -0.1640786$

$f_2 = -0.1377674$

$f_2 > f_1$

New Upper Bound = 2.472136

New Lower Bound = 2

New Upper Test Point = 2.291796

New Lower Test Point = 2.18034

Iteration # 4

$f_1 = -0.1315562$

$f_2 = -0.1640786$

$f_2 < f_1$

New Upper Bound = 2.472136

New Lower Bound = 2.18034

New Lower Test Point = 2.291796

New Upper Test Point = 2.36068

Iteration # 5

$f_1 = -0.1640786$

$f_2 = -0.1655449$

$f_2 < f_1$

New Upper Bound = 2.472136

New Lower Bound = 2.291796

New Lower Test Point = 2.36068

New Upper Test Point = 2.403252

Iteration # 6

$f_1 = -0.1655449$

$f_2 = -0.1593337$

$f_2 > f_1$

New Upper Bound = 2.403252

New Lower Bound = 2.291796

New Upper Test Point = 2.36068

New Lower Test Point = 2.334369

Iteration # 7

$f_1 = -0.1666651$

$f_2 = -0.1655449$

$f_2 > f_1$

New Upper Bound = 2.36068

New Lower Bound = 2.291796

New Upper Test Point = 2.334369

New Lower Test Point = 2.318107

Iteration # 8

$f_1 = -0.1663189$

$f_2 = -0.1666651$

$f_2 < f_1$

New Upper Bound = 2.36068

New Lower Bound = 2.318107

New Lower Test Point = 2.334369

New Upper Test Point = 2.344419

We accept our x value as the midpoint of 2.334369 and 2.344419. This value is 2.339394.

We back solve to find y, since the solution falls on the line, $3/2\, x + y = 7$. We compute y as $y = 3.4909$.

Our Nash arbitration point, using R, is $(2.339394, 3.4909)$. The value of the function $f(2.339394, 3.4909) = 0.16666$ (remember to change the sign if you take the functional value directly from R).

We still solve the following simultaneous equations:

$$2p_1 + 4p_2 = 2.3333$$

$$4p_1 + p_2 = 3.5$$

We find that the probabilities to be played are 5/6 and 1/6.

12.5 EXERCISES

12.1. Market strategies for companies A and B in the payoff matrix are given in Table 12.10. Determine the total conflict solution.

TABLE 12.10   Payoff Matrix for Exercise 12.1

		Company B	
		Locate in a town	**Locate in a small town**
Company A	Locate in a large town	(68, −68)	(50, −50)
	Locate in a small town	(45, −45)	(35, −35)

12.2. Market strategies for companies A and B in the payoff matrix are given in Table 12.11. Determine the total conflict solution.

TABLE 12.11   Payoff Matrix for Exercise 12.2

		Company B	
		Locate in a town	**Locate in a small town**
Company A	Locate in a large town	(68, 32)	(50, 50)
	Locate in a small town	(45, 55)	(35, 65)

12.3. Consider a hitter–pitcher duel with entries as provided in the payoff matrix (as given in Table 12.12). What should the hitter and pitcher do to obtain the best overall result?

TABLE 12.12   Hitter/Pitcher Payoff Matrix for Exercise 12.3

Hitter/Pitcher		Pitch		
Guess	**P**	**FB**	**Split**	**Curve**
FB		0.395	0.225	0.200
Split		0.250	0.260	0.195
Curve		0.200	0.225	0.273

12.4. Consider a change to the writer's guild example. The payoff matrix (Table 12.13) is the following partial conflict game. Determine what each player should do.

TABLE 12.13   Writer's Guild Payoff Matrix for Exercise 12.4

		Colin	
		C1	**C2**
Rose	R1	(0, 1)	(6, 2)
	R2	(3, 9)	(1, 0)

REFERENCES AND ADDITIONAL READINGS

Barron, E.N. (2013). *Game Theory: An Introduction*. New York: Wiley.

Fox, W.P. (2010). Teaching the applications of optimization in game theory's zero-sum and non-zero sum games. *International Journal of Data Analysis Techniques and Strategies (IDATS)* 2 (3): 258–284.

Giordano, F., W. Fox and S. Horton (2014). *A First Course in Mathematical Modeling*. Boston: Cengage Publishers.

Nash, John (1950). Equilibrium points in n-person games. *Proceedings of the National Academy of Sciences of the United States of America* 36 (1): 48–49.

Straffin, Philip D. (2004). *Game Theory and Strategy*. Washington: Mathematical Association of America.

Index

Note: Page numbers in *italic* indicate a figure and **bold** indicate a table on the corresponding page.

printed in the United States
by Baker & Taylor Publisher Services

Printed in the United States
by Baker & Taylor Publisher Services